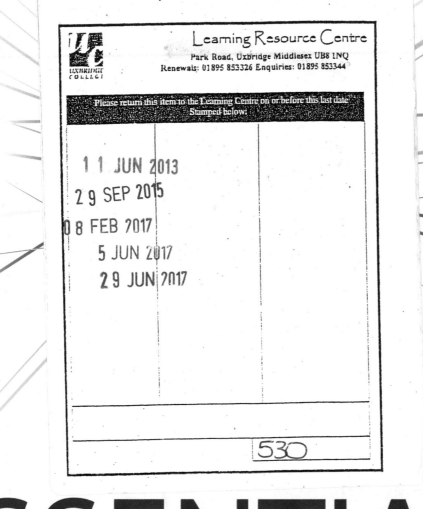
# ESSENTIALS

## AQA

## GCSE Physics

### Revision Guide

# Contents

# Contents

N.B. The numbers in brackets correspond to the reference numbers on the AQA GCSE Physics specification.

# How to Use This Guide

This revision guide has been written and developed to help you get the most out of your revision.

This guide covers both Foundation and Higher Tier content.

(HT) Content that will only be tested on the Higher Tier papers appears in a pale yellow tinted box labelled with the (HT) symbol.

- The **coloured page headers** clearly identify the separate units, so that you can revise for each one separately: Unit 1 is red, Unit 2 is purple, and Unit 3 is blue.
- The exam will include questions on **How Science Works**, so make sure you work through the How Science Works section in green at the front of this guide before each exam.
- There are two **summary pages** at the end of each unit, which outline all the key points. These are great for a final recap before your exam.

- There are **practice questions** at the end of each unit so you can test yourself on what you've just learned. (The answers are given on pages 83–84 so you can mark your own answers.)
- You'll find **key words** in a yellow box on each 2-page spread. They are also highlighted in colour within the text; higher tier key words are highlighted in a different colour. Make sure you know and understand all these words before moving on!
- There's a **glossary** at the back of the book. It contains all the key words from throughout the book so you can check any definitions you're unsure of.
- The **tick boxes** on the contents page let you track your revision progress: simply put a tick in the box next to each topic when you're confident that you know it.
- Don't just read the guide, **learn actively**! Constantly test yourself without looking at the text.

Good luck with your exam!

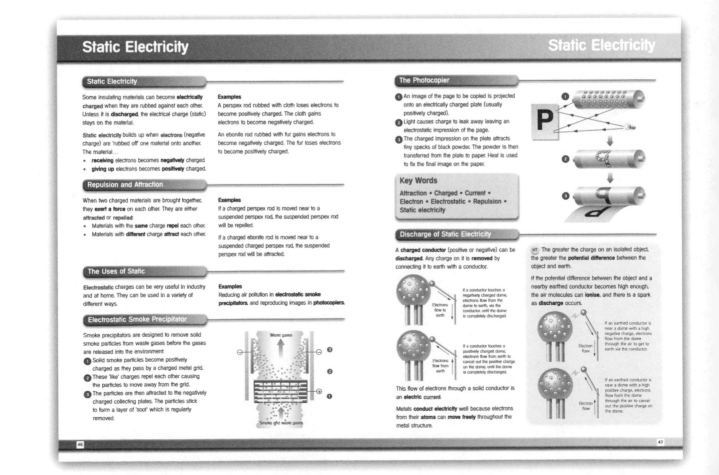

## How Science Works – Explanation

The AQA GCSE science specifications incorporate:

**1** **Science Content** – all the scientific explanations and evidence that you need to know for the exam. (It is covered on pages 12–79 of this revision guide.)

**2** **How Science Works** – a set of key concepts, relevant to all areas of science. It covers…

- the relationship between scientific evidence, and scientific explanations and theories
- how scientific evidence is collected
- how reliable and valid scientific evidence is
- the role of science in society
- the impact science has on our lives
- how decisions are made about the ways science and technology are used in different situations, and the factors affecting these decisions.

Your teacher(s) will have taught these two types of content together in your science lessons. Likewise, the questions on your exam papers will probably combine elements from both types of content. So, to answer the questions, you'll need to recall the relevant scientific facts *and* apply your knowledge of how science works.

The key concepts of How Science Works are summarised in this section of the revision guide (pages 6–11).

You should be familiar with all of these concepts. If there is anything you are unsure about, ask your teacher to explain it to you.

How Science Works is designed to help you learn about and understand the practical side of science. It aims to help you develop your skills when it comes to…

- evaluating information
- developing arguments
- drawing conclusions.

*N.B. Practical tips on how to evaluate information are included on page 11.*

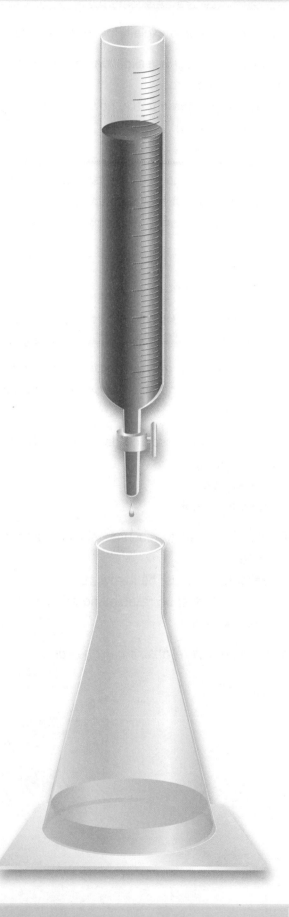

# How Science Works

## What is the Purpose of Science?

Science attempts to explain the world we live in.

Scientists carry out investigations and collect evidence in order to...
- **explain phenomena** (i.e. how and why things happen)
- **solve problems**.

Scientific knowledge and understanding can lead to the **development of new technologies** (e.g. in medicine and industry), which have a huge impact on...
- society
- the environment.

## What is the Purpose of Evidence?

Scientific evidence provides **facts** which answer a specific question and either **support** or **disprove** an idea / theory.

Evidence is often based on data that has been collected through...
- **observations**
- **measurements**.

To allow scientists to reach conclusions, evidence must be...
- **reliable** – it must be trustworthy
- **valid** – it must be reliable and answer the question.

*N.B. If data isn't reliable, it can't be valid.*

To ensure scientific evidence is reliable and valid, scientists use ideas and practices relating to...
1. observations
2. investigations
3. measurements
4. data presentation
5. conclusions.

These five key ideas are covered in more detail on the following pages.

## Observations

Most scientific investigations begin with an **observation**. A scientist observes an event / phenomenon and decides to find out more about how and why it happens.

The first step is to develop a **hypothesis** to suggest an explanation for the phenomenon. Hypotheses normally suggest a relationship between two or more **variables** (factors that change). Hypotheses are based on…

- careful observations
- existing scientific knowledge
- a bit of creative thinking.

The hypothesis is used to make a **prediction**, which can be tested through scientific investigation. The data collected from the investigation will…

- support the hypothesis **or**
- show it to be untrue **or**
- lead to the development of a new hypothesis.

If new observations or data don't match existing explanations or theories, they must be checked for reliability and validity.

Sometimes, if the new observations and data are valid, existing theories and explanations have to be revised or amended, and so scientific knowledge grows and develops.

**Example**

1. A scientist **observes** that freshwater shrimp are only found in certain parts of a stream.

2. He uses scientific knowledge of shrimp and water flow to develop a **hypothesis**, which relates the presence of shrimp (first variable) to the rate of water flow (second variable).

3. He **predicts** that shrimp are only found in parts of the stream where the water-flow rate is below a certain value.

4. He **investigates** by recording the presence of shrimp in different parts of the stream, where flow rates differ.

5. His **data** shows that shrimp are only present in parts of the stream where the flow rate is below a certain value. So, the data **supports** the hypothesis. But, it also shows that shrimp aren't *always* present in these parts of the stream.

6. The scientist realises there must be another factor affecting the distribution of shrimp. He **refines his hypothesis**.

# How Science Works

An **investigation** involves collecting data to find out whether there is a relationship between two **variables**. A variable is a factor that can take different values.

In an investigation there are two variables:

1. **Independent variable** – can be adjusted (changed) by the person carrying out the investigation.
2. **Dependent variable** – measured each time a change is made to the independent variable, to see if it also changes.

Variables can have different types of values:

- **Continuous variables** – take numerical values. These are usually measurements, e.g. temperature.
- **Discrete variables** – only take whole-number values. These are usually quantities, e.g. the number of shrimp in a stream.
- **Ordered variables** – have relative values, e.g. 'small', 'medium' or 'large'.
- **Categoric variables** – have a limited number of specific values, e.g. different breeds of dog.

*N.B. Numerical values tend to be more informative than ordered and categoric variables.*

An investigation tries to find out whether an **observed** link between two variables is...

- **causal** – a change in one variable causes a change in the other
- **due to association** – the changes in the two variables are linked by a third variable
- **due to chance** – the change in the two variables is unrelated; it is coincidental.

### Fair Tests

In a **fair test**, the only factor that can affect the dependent variable is the independent variable. Other **outside variables** that could influence the results are kept the same or eliminated.

It's a lot easier to carry out a fair test in the lab than in the field, where conditions can't always be controlled. The impact of an outside variable, like the weather, has to be reduced by ensuring all measurements are affected by it in the same way.

### Accuracy and Precision

How accurate the data collected needs to be depends on what the investigation is trying to find out. For example, measures of alcohol in the blood must be accurate to determine whether a person is legally fit to drive.

The data collected must be **precise** enough to form a **valid conclusion**: it should provide clear evidence for or against the hypothesis.

To ensure data is as accurate as possible, you can…

- calculate the **mean** (average) of a set of repeated measurements to get a **best estimate** of the true value
- increase the number of measurements taken to improve the **accuracy** and the **reliability** of the mean.

## Measurements

Apart from outside variables, there are a number of factors that can affect the reliability and validity of measurements:

- **Accuracy of instruments** – depends on how accurately the instrument has been calibrated. (Expensive equipment is usually more accurately calibrated.)
- **Sensitivity of instruments** – determined by the smallest change in value that the instrument can detect. For example, bathroom scales aren't sensitive enough to detect the changes in a baby's weight, but the scales used by a midwife are.

- **Human error** – can occur if you lose concentration. Systematic (repeated) errors can occur if the instrument hasn't been calibrated properly or is misused.

You need to examine any **anomalous** (irregular) values to try to determine why they appear. If they have been caused by an equipment failure or human error, it is common practice to discount them from any calculations.

## Presenting Data

Data is often presented in a **chart** or **graph** because it makes…
- the patterns more evident
- it easier to see the relationship between two variables.

The relationship between variables can be…
- **linear** (positive or negative)
- **or directly proportional**.

If you present data clearly, it is easier to identify any anomalous values. The type of chart or graph you use to present data depends on the type of variable involved:

1. **Tables** organise data (but patterns and anomalies aren't always obvious).
2. **Bar charts** display data when the independent variable is categoric or discrete and the dependent variable is continuous.
3. **Line graphs** display data when both variables are continuous.
4. **Scattergrams** (scatter diagrams) show the underlying relationship between two variables. This can be made clearer if you include a **line of best fit**.

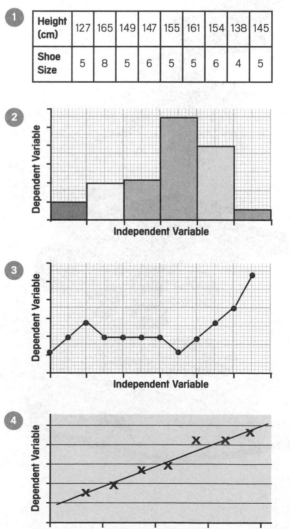

| Height (cm) | 127 | 165 | 149 | 147 | 155 | 161 | 154 | 138 | 145 |
|---|---|---|---|---|---|---|---|---|---|
| Shoe Size | 5 | 8 | 5 | 6 | 5 | 5 | 6 | 4 | 5 |

# How Science Works

Conclusions **should**...
- describe the patterns and relationships between variables
- take all the data into account
- make direct reference to the original hypothesis / prediction.

Conclusions **shouldn't**...
- be influenced by anything other than the data collected
- disregard any data (except anomalous values)
- include any speculation.

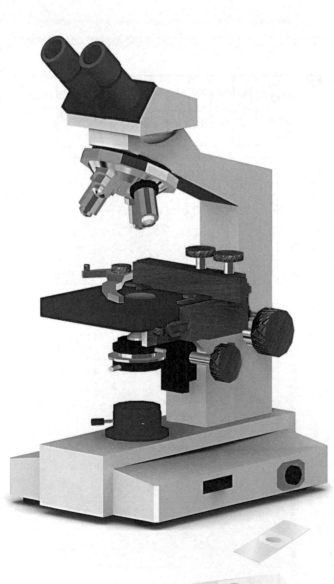

**Evaluation**

An **evaluation** looks at the whole investigation. It should consider...
- the original purpose of the investigation
- the appropriateness of the methods and techniques used
- the reliability and validity of the data
- the validity of the conclusions.

The **reliability** of an investigation can be increased by...
- looking at relevant data from secondary sources
- using an alternative method to check results
- ensuring that the results can be reproduced by others.

**Science and Society**

Scientific understanding can lead to technological developments. These developments can be exploited by different groups of people for different reasons. For example, the successful development of a new drug...
- benefits the drugs company financially
- improves the quality of life for patients.

Scientific developments can raise certain **issues**. An issue is an important question that is in dispute and needs to be settled. The resolution of an issue may not be based on scientific evidence alone.

There are several different **issues** which can arise:
- **Social** – the impact on the human population of a community, city, country, or the world.
- **Economic** – money and related factors like employment and the distribution of resources.
- **Environmental** – the impact on the planet, its natural ecosystems and resources.
- **Ethical** – what is morally right and wrong; requires a valued judgement to be made about what is acceptable.

*N.B. There is often an overlap between social and economic issues.*

## Evaluating Information

It is important to be able to evaluate information relating to social-scientific issues, both for the exam and to help you make informed decisions in life.

When evaluating information...
- make a list of **pluses**
- make a list of **minuses**
- consider how each point might **impact on society**.

*N.B. Remember,* **PMI** *– pluses, minuses, impact on society.*

You also need to consider whether the source of information is reliable and credible. Some Important factors to consider are...
- **opinions**
- **bias**
- **weight of evidence**.

**Opinions** are personal viewpoints. Opinions backed up by valid and reliable evidence carry far more weight than those based on non-scientific ideas.

Information is **biased** if it favours one particular viewpoint without providing a balanced account. Biased information might include incomplete evidence or try to influence how you interpret the evidence.

Scientific evidence can be given **undue weight** or dismissed too lightly due to...
- political significance
- status (academic or professional status, experience, authority and reputation).

### Limitations of Science

Although science can help us in lots of ways, it can't supply all the answers. We are still finding out about things and developing our scientific knowledge.

There are some questions that science can't answer. These tend to be questions relating to...
- ethical issues
- situations where it isn't possible to collect reliable and valid scientific evidence.

Science can often tell us if something **can** be done, and **how** it should be done, but it can't tell us whether it **should** be done.

# Conduction, Convection and Radiation

## Conduction

**Conduction** is the **transfer** of heat **energy** without the substance itself moving.

The structure of **metals** makes them good **conductors** of heat:

**1** As a metal becomes hotter, its tightly packed particles gain more kinetic energy and vibrate.

**2** This energy is transferred to cooler parts of the metal by delocalised electrons, which move freely through the metal, colliding with particles and other electrons.

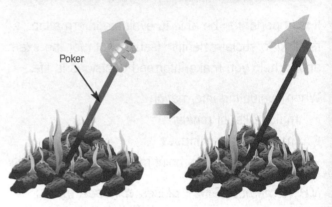

Heat energy is conducted up the poker as the hotter parts transfer energy to the colder parts

## Convection

**Convection** is the transfer of heat energy through **movement**.

Convection occurs in **liquids** and **gases** and creates **convection currents**:

**1** The particles in the liquid or gas which are nearest the heat source move faster causing the substance to expand and become less dense than in the colder parts.

**2** The warm liquid or gas will rise up. The liquid or gas cools, becomes denser and sinks. The colder, denser liquid or gas moves into the space created (close to the heat source), replacing the liquid or gas that has risen.

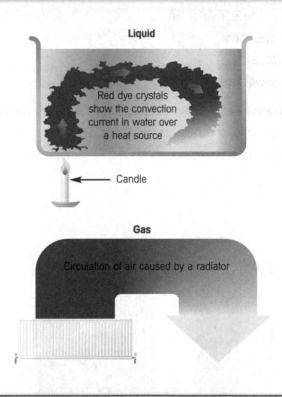

**Liquid**

Red dye crystals show the convection current in water over a heat source

Candle

**Gas**

Circulation of air caused by a radiator

## Radiation

**Thermal (infra red) radiation** is the transfer of heat (thermal) energy by **electromagnetic waves**. No particles of matter are involved.

- All objects emit and absorb thermal radiation.
- The hotter the object, the more energy it radiates.
- The amount of radiation an object gives out or takes in depends on its **surface**, **shape** and **dimensions**.

An object will emit or absorb energy faster if there's a big difference in temperature between it and its surroundings.

Under similar conditions, different materials transfer heat at different rates. At the same temperature, dark matt surfaces...

- emit more radiation than light shiny surfaces
- absorb more radiation than light shiny surfaces.

## Transferring and Transforming Energy

When devices transfer energy, **only part** of the energy is usefully transferred to **where** it's wanted and in **the form** that's wanted.

The remaining energy is **transformed** in a non-useful way, mainly as heat energy. It is known as **wasted energy**.

For example, a light bulb transforms electrical energy into useful light energy. However, most of the energy is wasted as heat energy.

The wasted energy and the useful energy are eventually transferred to their surroundings, which become warmer.

**No energy is created or destroyed. It is just changed into a different form (transformed).**

But, the energy becomes increasingly spread out, so it's difficult for any further useful energy transfers to occur.

## Efficiency

The **efficiency** of a device refers to the proportion of energy that is usefully transformed. The greater the proportion of energy that is usefully transformed, the more **efficient** the device is.

For example, only a quarter of the energy supplied to a television is usefully transformed into light and sound. So it's only 25% efficient.

Electrical energy
200 joules/s

Wasted energy
Heat 150 joules/s

Useful energy
Light 20 joules/s

Useful energy
Sound 30 joules/s

$$\text{Efficiency (\%)} = \frac{\text{Useful energy transferred by device}}{\text{Total energy supplied to device}} \times 100$$

## Key Words

**Conduction • Conductor • Convection • Efficiency • Energy • Radiation • Transfer • Transform**

# Energy Transformation and Transfer

## Energy Transformation

Most of the **energy** transferred to homes and industry is **electrical energy**.

Electrical energy is easily transformed into…
- **heat / thermal energy** (e.g. hairdryer)
- **light energy** (e.g. lamp)
- **sound energy** (e.g. stereo speakers)
- **kinetic / movement energy** (e.g. electric fan).

The amount of energy transformed by an electrical appliance depends on…
- how long the appliance is switched on
- how fast the appliance can transform energy.

The **power** of an appliance is measured in **watts** (W) or **kilowatts** (kW).

**Energy** is normally measured in **joules** (J).

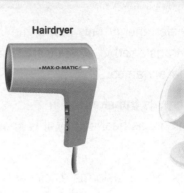

Hairdryer

Lamp

Stereo Speakers

Electric Fan

## Energy Transfer

Electricity is generated at power stations. It is then transferred to homes, schools and factories by a network of cables called the **National Grid**.

**Transformers** are used to change the **voltage** of the **alternating current** supply, before and after it is transmitted through the National Grid. **Step-up** and **step-down transformers** are used.

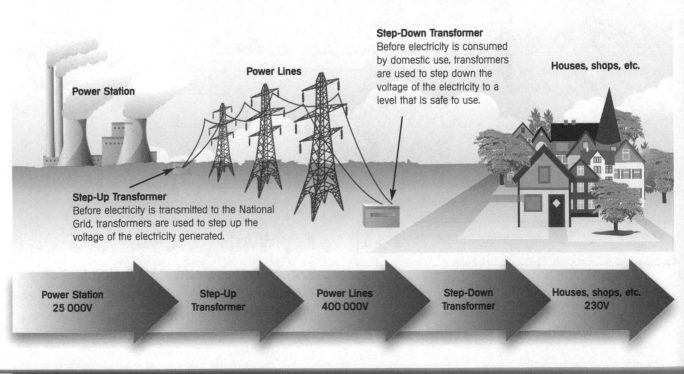

**Step-Down Transformer**
Before electricity is consumed by domestic use, transformers are used to step down the voltage of the electricity to a level that is safe to use.

Houses, shops, etc.

Power Lines

Power Station

**Step-Up Transformer**
Before electricity is transmitted to the National Grid, transformers are used to step up the voltage of the electricity generated.

| Power Station 25 000V | Step-Up Transformer | Power Lines 400 000V | Step-Down Transformer | Houses, shops, etc. 230V |

# Energy Transformation and Transfer

## Reducing Energy Loss

The higher the **current** that passes through a wire, the greater the amount of energy that is **lost as heat** from the wire. So, as low a current as possible needs to be transmitted through the power lines. Increasing voltage (potential difference) lowers the current.

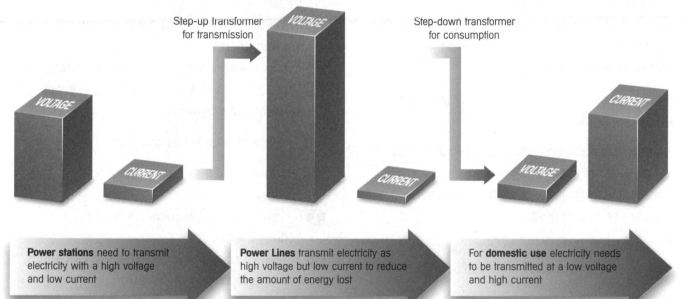

Step-up transformer for transmission

Step-down transformer for consumption

**Power stations** need to transmit electricity with a high voltage and low current

**Power Lines** transmit electricity as high voltage but low current to reduce the amount of energy lost

For **domestic use** electricity needs to be transmitted at a low voltage and high current

## Energy Calculations

You can calculate the **amount of energy transferred** from the mains using:

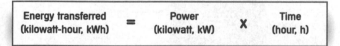

| Energy transferred (kilowatt-hour, kWh) | = | Power (kilowatt, kW) | x | Time (hour, h) |
|---|---|---|---|---|

You can calculate the **cost of energy transferred** from the mains using:

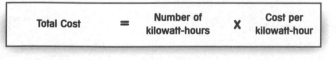

| Total Cost | = | Number of kilowatt-hours | x | Cost per kilowatt-hour |
|---|---|---|---|---|

### Key Words

**Current • Energy • National Grid • Power • Thermal energy • Transformer • Voltage**

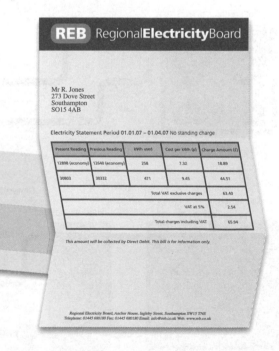

**REB** Regional**Electricity**Board

Mr R. Jones
273 Dove Street
Southampton
SO15 4AB

Electricity Statement Period 01.01.07 – 01.04.07 No standing charge

| Present Reading | Previous Reading | kWh used | Cost per kWh (p) | Charge Amount (£) |
|---|---|---|---|---|
| 12898 (economy) | 12640 (economy) | 258 | 7.32 | 18.89 |
| 30803 | 30332 | 471 | 9.45 | 44.51 |
| | | | Total VAT exclusive charges | 63.40 |
| | | | VAT at 5% | 2.54 |
| | | | Total charges including VAT | 65.94 |

This amount will be collected by Direct Debit. This bill is for information only.

Regional Electricity Board, Anchor House, Ingleby Street, Southampton SW15 7NE
Telephone: 01445 680180 Fax: 01445 680180 Email: info@reb.co.uk Web: www.reb.co.uk

# Energy Sources

## Fuels

**Fuels** are substances which release useful amounts of **energy** when they are burned.

## Non-Renewable Energy Sources

**Coal, oil** and **gas** are **fossil fuels**. They are **energy sources** which we depend on for most of the energy that we use. They can't be replaced within a lifetime, so they will eventually run out. They are **non-renewable energy sources**.

**Nuclear fuels** such as **uranium** and **plutonium** are also **non-renewable**.

**Nuclear fission** is the release of nuclear particles that collide with the nuclei of other atoms, splitting them. This causes a chain reaction that generates huge amounts of heat energy.

Nuclear fuel isn't burned like coal, oil or gas to release energy and it isn't classed as a fossil fuel.

Non-renewable energy sources can be used to generate electricity.

Fossil fuels (and wood) generate electricity in power stations:

1 Fossil fuels (and wood) are burned to release **thermal energy**.

2 The thermal energy boils water to produce steam.

3 The steam drives turbines which are attached to an electrical generator.

*N.B. Wood isn't a fossil fuel and it isn't non-renewable.*

Nuclear fuel is used to generate electricity in a similar way:

1 A reactor is used to generate heat by nuclear fission.

2 A heat exchanger transfers the thermal energy from the reactor to the water.

3 The water turns to steam and drives the turbines.

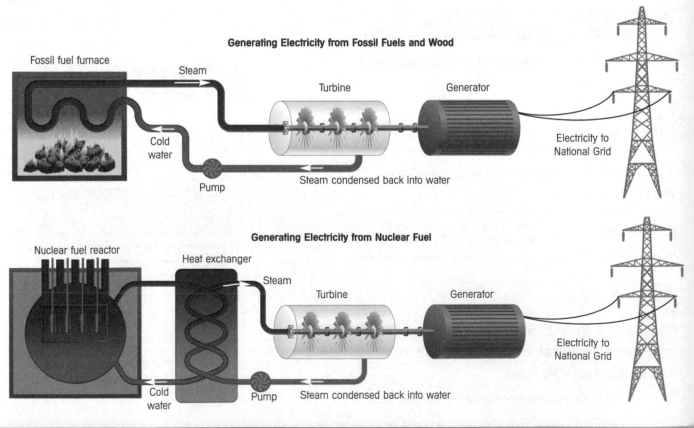

**Generating Electricity from Fossil Fuels and Wood**

Fossil fuel furnace — Steam — Turbine — Generator — Electricity to National Grid — Cold water — Steam condensed back into water — Pump

**Generating Electricity from Nuclear Fuel**

Nuclear fuel reactor — Heat exchanger — Steam — Turbine — Generator — Electricity to National Grid — Cold water — Pump — Steam condensed back into water

## Non-Renewable Energy Sources

| Source | Advantages | Disadvantages |
|---|---|---|
| Coal | • Relatively cheap and easy to obtain.<br>• Coal-fired power stations have a relatively quick start-up time.<br>• There may be over a century's worth of coal left. | • Burning produces $CO_2$ and $SO_2$.<br>• Produces more $CO_2$ per unit of energy than oil or gas does.<br>• $SO_2$ causes **acid rain**. (Removing the $SO_2$ is costly.) |
| Oil | • Relatively easy to find, though the price is variable.<br>• Oil-fired power stations are flexible in meeting demand.<br>• Oil-fired power stations have a quick start-up time.<br>• There is enough oil left for the short–medium term. | • Burning produces $CO_2$ and $SO_2$.<br>• Produces more $CO_2$ per unit of energy than gas does.<br>• Tankers pose the risk of **spillage** and **pollution**. |
| Gas | • Can be found as easily as oil.<br>• Gas-fired power stations have the quickest start-up time.<br>• There is enough gas left for the short–medium term.<br>• Doesn't produce $SO_2$. | • Burning produces $CO_2$ (although it produces less per unit of energy than coal or oil).<br>• Expensive pipelines and networks are often required to transport it. |
| Nuclear | • Cost and rate of fuel is relatively low.<br>• Can be situated in sparsely populated areas.<br>• Nuclear power stations are flexible in meeting demand.<br>• Doesn't produce $CO_2$ or $SO_2$. | • Although there is very little escape of radioactive material in normal use, radioactive waste can stay **dangerously radioactive** for thousands of years.<br>• Building and decommissioning is costly.<br>• Longest start-up time. |

## Key Words

**Energy • Fossil fuel • Non-renewable • Nuclear • Thermal energy**

# Energy Sources

## Renewable Energy Sources

**Renewable energy sources** will not run out; they are continually being replaced.

Many renewable energy sources are 'powered' by the **Sun** or **Moon**.

The gravitational pull of the Moon creates tides.

The Sun causes…

- evaporation, which results in rain and flowing water
- convection currents, which result in winds, which create waves.

Renewable energy sources can be used to drive turbines or generators directly.

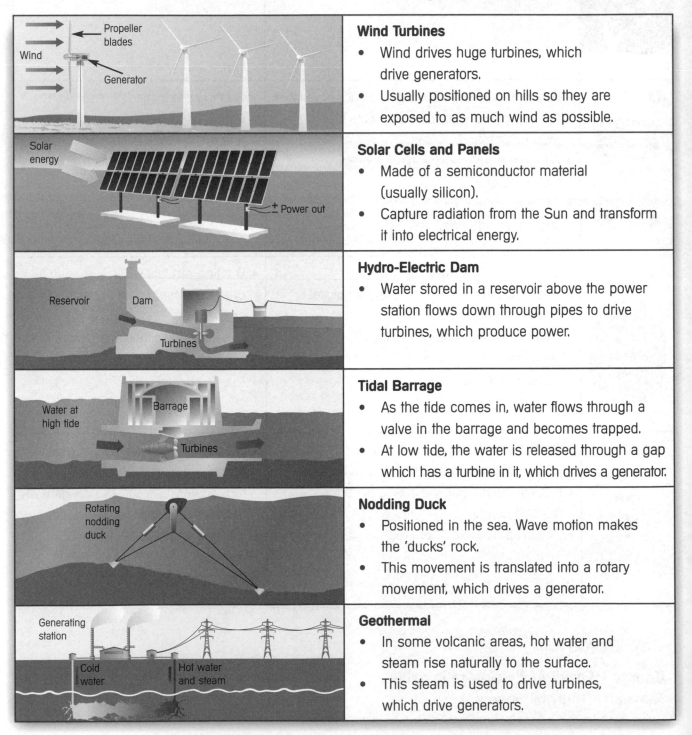

### Wind Turbines
- Wind drives huge turbines, which drive generators.
- Usually positioned on hills so they are exposed to as much wind as possible.

### Solar Cells and Panels
- Made of a semiconductor material (usually silicon).
- Capture radiation from the Sun and transform it into electrical energy.

### Hydro-Electric Dam
- Water stored in a reservoir above the power station flows down through pipes to drive turbines, which produce power.

### Tidal Barrage
- As the tide comes in, water flows through a valve in the barrage and becomes trapped.
- At low tide, the water is released through a gap which has a turbine in it, which drives a generator.

### Nodding Duck
- Positioned in the sea. Wave motion makes the 'ducks' rock.
- This movement is translated into a rotary movement, which drives a generator.

### Geothermal
- In some volcanic areas, hot water and steam rise naturally to the surface.
- This steam is used to drive turbines, which drive generators.

## Renewable Energy Sources

The energy sources listed below can provide clean, safe energy. Some of their advantages and disadvantages are given.

| Source | Advantages | Disadvantages |
|---|---|---|
| **Wind** | • No fuel and little maintenance required.<br>• No pollutant gases produced.<br>• Can be built offshore. | • Cause noise and visual pollution.<br>• Not very flexible in meeting demand (unless energy is stored).<br>• High capital outlay to build them. |
| **Tidal and Waves** | • No fuel required.<br>• No pollutant gases produced.<br>• Barrage water can be released when electricity demand is high. | • They are unsightly, a hazard to shipping, and destroy habitats.<br>• Variations of tides and waves affect output.<br>• High capital outlay to build them. |
| **Hydro-Electric** | • Fast start-up time.<br>• No pollutant gases produced.<br>• Water can be pumped back to the reservoir when electricity demand is low. | • Location is critical: often involves damming upland valleys.<br>• There must be adequate rainfall in the region where the reservoir is.<br>• Very high initial capital outlay. |
| **Solar** | • Can produce electricity in remote locations.<br>• No pollutant gases produced. | • Dependent on intensity of light.<br>• High cost per unit of electricity produced. |

## Summary

| Advantages | Disadvantages |
|---|---|
| • No fuel costs during operation.<br>• No pollution.<br>• Often low maintenance. | • Produce small amounts of electricity (except hydro-electric).<br>• Take up lots of space and are unsightly.<br>• Unreliable (apart from hydro-electric) – they depend on the weather.<br>• High initial capital outlay. |

## Key Words

**Energy • Renewable**

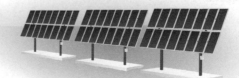

# Unit 1a Summary

## Conduction and Convection

Conduction = transfer of heat energy without the substance moving (solids).

Convection = transfer of heat energy through movement (gases and liquids).

## Radiation

Thermal radiation = transfer of heat energy by **electromagnetic waves**.

Dark matt surfaces emit more radiation than light shiny surfaces.
Dark matt surfaces absorb more radiation than light shiny surfaces.

## Transferring and Transforming Energy

When devices transfer energy, only some of it is usefully transferred to where it is wanted and in the form that is wanted. The rest is **wasted energy**.

No energy is created or destroyed. It is just changed into a different form.

**Efficiency** = the proportion of usefully transformed energy in a device.

Electrical energy can be transformed into…
- heat energy
- light energy
- sound energy
- kinetic energy.

The amount of energy transformed by an electrical appliance depends on…
- how long the appliance is switched on
- how fast it can transform energy.

Power = measured in watts (W) / kilowatts (kW)

Energy = measured in joules (J)

## National Grid

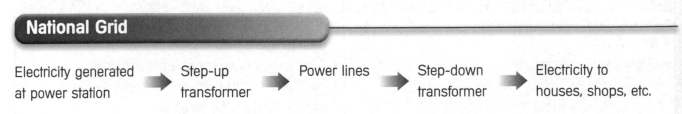

Electricity generated at power station → Step-up transformer → Power lines → Step-down transformer → Electricity to houses, shops, etc.

The higher the current that passes through a wire, the more energy is lost as heat. So, as low a current as possible needs to pass through power lines. Increasing voltage lowers current.

## Energy Calculations

| Energy transferred (kilowatt-hour, kWh) | = | Power (kilowatt, kW) | x | Time (hour, h) |
|---|---|---|---|---|

| Cost of energy transferred | = | Number of kilowatt-hours | x | Cost per kilowatt-hour |
|---|---|---|---|---|

## Non-Renewable Energy Sources

Non-renewable energy sources = **fossil fuels** (coal, oil, gas) and **nuclear fuels**.

Nuclear fission = the release of nuclear particles that collide with the nuclei of other atoms, splitting them, and causing a chain reaction that generates large amounts of heat energy.

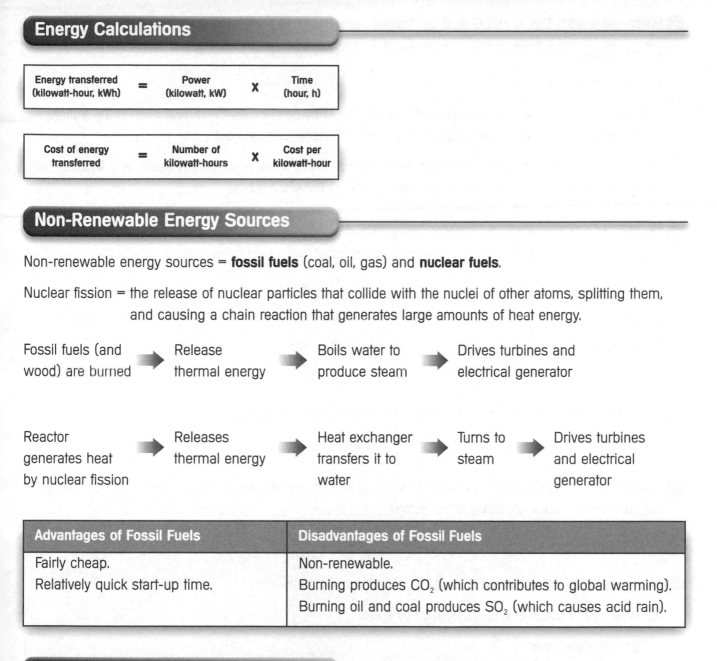

Fossil fuels (and wood) are burned → Release thermal energy → Boils water to produce steam → Drives turbines and electrical generator

Reactor generates heat by nuclear fission → Releases thermal energy → Heat exchanger transfers it to water → Turns to steam → Drives turbines and electrical generator

| Advantages of Fossil Fuels | Disadvantages of Fossil Fuels |
|---|---|
| Fairly cheap. Relatively quick start-up time. | Non-renewable. Burning produces $CO_2$ (which contributes to global warming). Burning oil and coal produces $SO_2$ (which causes acid rain). |

## Renewable Energy Sources

Renewable energy sources = wind turbines, Solar cells / panels, Hydro-electric dams, Tidal barrage, Nodding duck, Geothermal.

Many are powered by the Sun or Moon.

| Advantages of Renewable Energy Sources | Disadvantages of Renewable Energy Sources |
|---|---|
| No fuel costs. No pollution. Usually low maintenance. | Produce small amounts of electricity. High initial set-up costs. Unsightly. |

# Unit 1a Practice Questions

**1** Match the words A, B, C and D with the spaces numbered 1 to 4 in the sentences below.

**A** transferred ............................................ **B** convection ............................................

**C** electromagnetic ............................................ **D** conduction ............................................

Heat can be ___**1**___ from a hotter place to colder places by different methods. Radiation is the transfer of heat energy as an ___**2**___ wave, where no particles of matter are involved.

When heat energy is transferred without the substance moving this is ___**3**___. In gases and liquids, heat energy is transferred by movement. This type of energy transfer is called ___**4**___.

**2** The four devices below transfer electrical energy into different forms of energy. Match each of the devices A, B, C and D with the word numbered 1 to 4 that best describes the useful energy transfer.

**A** torch ............................................ **B** radio ............................................

**C** drill ............................................ **D** electric fire ............................................

**1** sound **2** heat

**3** kinetic **4** light

**3** How is electricity transferred from power stations to homes, schools, etc.?

............................................................................................................................................

**4** A kettle takes in 2000 joules/s of energy which is usefully transferred as 1500 joules/s of heat energy in the water. What is the efficiency of the kettle?

............................................................................................................................................

**5** This question has four parts. For each part, tick the correct answer.

**a)** What is the useful energy transferred by a laptop?

   **i)** Light and heat

   **ii)** Sound and light

   **iii)** Sound and heat

   **iv)** Movement and heat

**b)** What is the wasted energy transferred by a laptop?

   **i)** Light

   **ii)** Sound

   **iii)** Heat

   **iv)** Movement and heat

**c)** A laptop takes in 1300 joules/s of energy which is usefully transferred as 900 joules/s of light and sound energy. How much energy is wasted?

   **i)**   400 joules/s as heat.

   **ii)**  400 joules/s as sound.

   **iii)** 399 joules as sound and heat.

   **iv)** 31% of the energy as sound and heat.

**d)** Which of these statements about the energy transfer in a laptop is false?

   **i)**   The energy is spread out to the environment.

   **ii)**  The energy is used up.

   **iii)** The energy warms up the surroundings.

   **iv)** It is difficult to use the energy.

**6** Speed warning signs use solar cells to power them. They transfer 45 joules of energy every second to charge batteries.

**a)** What form of energy does the solar cell use?

**b)** What is the maximum energy transferred to the battery in 5 minutes?

**c)** Give two advantages of using a solar cell for road warning signs.

**7 a)** Name two fossil fuels.

**b)** List three renewable energy sources.

**8** Match the energy sources A, B, C and D with the descriptions numbered 1 to 4 below.

**A** coal          **B** nuclear

**C** solar         **D** gas

**1** Expensive pipelines and networks needed to transport it.
**2** No pollutant gases produced.
**3** Expensive to build and decommission.
**4** Transforms chemical energy into heat energy.

# Electromagnetic Radiation

## Electromagnetic Radiation

**Electromagnetic radiations** are disturbances in an **electric field**.

They travel as **waves** and move **energy** from one place to another. Each type of electromagnetic radiation…
- has a different **wavelength**
- has a different **frequency**.

All types of electromagnetic radiation travel at the same speed through a **vacuum**.

Electromagnetic radiations form the **electromagnetic spectrum**.

Different wavelengths of electromagnetic radiation are **reflected**, **absorbed** or **transmitted** in different ways by different substances and types of surface.

When a wave (radiation) is absorbed by a substance…
- the energy is absorbed and makes the substance heat up
- it may create an alternating current of the same frequency as the radiation.

Electromagnetic waves obey the wave formula:

| Wave speed (metre/second, m/s) | = | Frequency (hertz, Hz) | x | Wavelength (metre, m) |
|---|---|---|---|---|

**Visible light** is one type of electromagnetic radiation. The seven 'colours of the rainbow' form the **visible spectrum** (the only part of the electromagnetic spectrum that can be seen).

The visible spectrum is produced because white light is made up of different colours. The colours are **refracted** by different amounts as they pass through a prism:
- **Red light is refracted the least.**
- **Violet light is refracted the most.**

### Key Words

**Absorption • Electromagnetic spectrum • Radiation • Reflection • Refraction • Transmission**

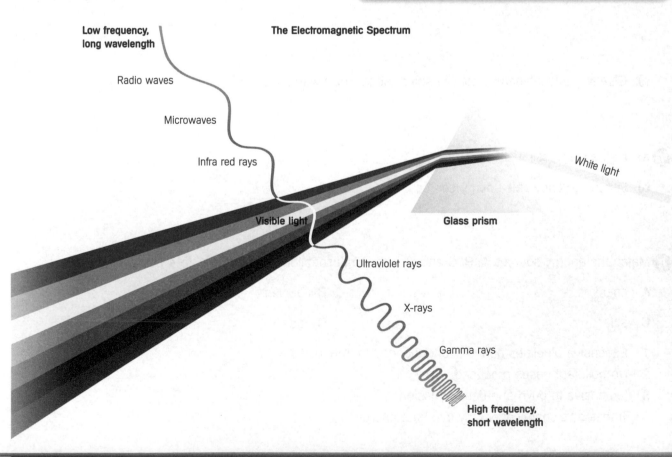

**The Electromagnetic Spectrum**

Low frequency, long wavelength

Radio waves

Microwaves

Infra red rays

Visible light

Ultraviolet rays

X-rays

Gamma rays

High frequency, short wavelength

White light

Glass prism

# Electromagnetic Radiation

| Electromagnetic Waves | Uses | Effects |
| --- | --- | --- |
| **Radio Waves** | • Transmitting radio and TV signals between places across the Earth. | • High levels of exposure for short periods can increase body temperature, leading to tissue damage. |
| **Microwaves** | • Satellite communication networks and mobile phone networks (they can pass through the Earth's atmosphere).<br>• Cooking – water molecules absorb microwaves, and heat up. | • May damage or kill cells because microwaves are absorbed by water in the cells, which heat up. |
| **Infra Red Rays** | • Grills, toasters and radiant heaters (e.g. electric fires).<br>• Remote controls for televisions, etc.<br>• **Optical fibre** communication. | • Absorbed by skin and felt as heat.<br>• An excessive amount can cause burns. |
| **Ultraviolet Rays** | • Security coding – special paint absorbs UV and emits visible light.<br>• Sun tanning and sunbeds. | • Passes through skin to the tissues below.<br>• High doses can kill cells.<br>• A low dose can cause **cancer**. |
| **X-Rays** | • Producing shadow pictures of bones and metals.<br>• Treating certain cancers. | • Passes through soft tissues (some is absorbed).<br>• High doses can kill cells.<br>• A low dose can cause cancer. |
| **Gamma Rays** | • Killing cancerous cells.<br>• Killing bacteria on food and surgical instruments. | • Passes through soft tissues (some is absorbed).<br>• High doses can kill cells.<br>• A low dose can cause cancer. |

# Communication

## Electromagnetic Wave Communication

**Sound**, e.g. speech or music, can be sent over long distances if it is converted into electrical signals which match the **frequency** and **amplitude** of the sound waves.

These electrical signals can be sent using...

- **cables** – copper cables weaken the signal during transmission, so regular amplification of the signal is required
- **electromagnetic waves** – a radio wave (called a 'carrier') carries the electrical signal from a transmitter. This produces a **modulated** wave.

The modulated wave is picked up by the aerial in a radio.

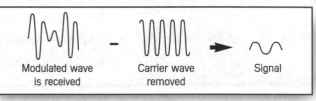

| Signal | Carrier | Modulated wave is transmitted |

It is then **demodulated** (i.e. the carrier wave is removed) to leave the original signal.

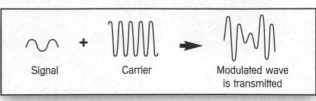

| Modulated wave is received | Carrier wave removed | Signal |

## Optical Fibres

Information can be transmitted using **optical fibres**. The electrical signal is converted into **visible light** or **infra red** pulses.

These waves can only travel in straight lines, but they reflect off the walls of the optical fibre to follow the curve of the fibre.

**Optical Fibre**

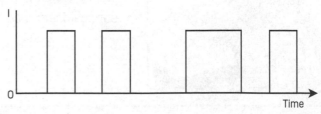

## Analogue and Digital Signals

**Analogue signals** vary continually in amplitude and / or frequency.

**Digital signals** don't vary. They have discrete values, usually **on (1)** or **off (0)**. The information is a **series of pulses**.

- Digital signals are of a better quality than analogue signals.
- There is no change in the signal information during transmission of digital signals.
- More information can be transmitted digitally in a given time by cable, optical fibre or carrier wave.
- Digital signals can be processed by computers.
- Analogue signals can pick up interference during transmission.

**Analogue Signals**

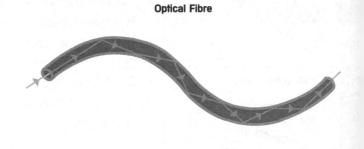

**Digital Signals**

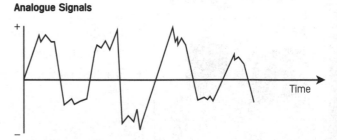

## Isotopes

An **atom** has a small central nucleus made up of **protons** and **neutrons**. The nucleus is surrounded by **electrons**.

All the atoms of a particular **element** have the same number of protons. However, they can have **different numbers of neutrons**. These are called **isotopes**.

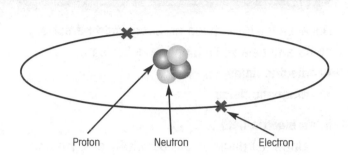

Proton    Neutron    Electron

## Radiation

Some substances give out **radiation** all the time. These substances are **radioactive**. Radiation is released from the nucleus of an atom as the result of a change in the atom's structure.

There are three types of radiation:

① **Alpha** (α) particles – a helium nucleus (made up of two protons and two neutrons).

② **Beta** (β) particles – a high-energy electron which is ejected from the nucleus.

③ **Gamma** (γ) rays – high-frequency electromagnetic radiation.

When radiation collides with atoms or molecules, it can knock electrons out of their structure, creating a **charged particle** called an **ion**.

Each type of radiation has a different…

- relative **ionising power**
- penetrating power
- range in air.

| Particle | Ionising Power | Penetrating Power |
|---|---|---|
| **Alpha** | Strong | Absorbed by a few centimetres of air or thin paper. |
| **Beta** | Reasonable | Passes through air and paper. Absorbed by a few centimetres of aluminium. |
| **Gamma** | Weak | Very penetrating. Needs many centimetres of lead or metres of concrete to stop it. |

### Key Words

**Analogue • Atom • Digital • Element • Ion • Ionising power • Isotope • Optical fibre • Radiation**

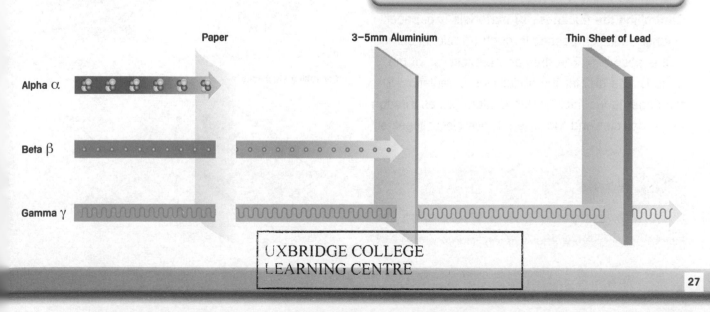

Paper          3–5mm Aluminium          Thin Sheet of Lead

Alpha α

Beta β

Gamma γ

# Radiation

## Electric and Magnetic Fields

Because they are made up of **charged particles**, alpha and beta **radiations** are deflected by…

- **electric fields**
- **magnetic fields**.

In an **electric field**…

- **alpha** (α) particles are **positively** charged and are deflected towards the negative electrode
- **beta** (β) particles are **negatively** charged and are deflected towards the **positive** electrode.
- **gamma** (γ) radiation is not deflected by the electric field.

In a **magnetic field**…

- **alpha** (α) particles are deflected.
- **beta** (β) particles are deflected.
- **gamma** (γ) radiation is not deflected by the magnetic field.

**Electric Field**

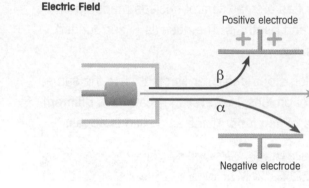

**Magnetic Field**

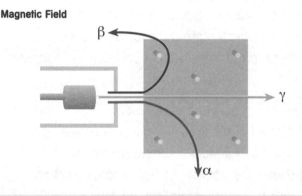

## Common Uses of Radiation

**Sterilisation** – gamma rays can be used to sterilise medical instruments and food.

**Treating cancer** – gamma radiation can be used to destroy cancerous cells.

**Tracers** – a tracer is a small amount of a radioisotope (radioactive isotope), which is put into a system. Its progress through the system can be traced using a radiation detector.

**Controlling the thickness of materials** (e.g. paper) – when radiation passes through a material some of it is absorbed. The greater the thickness of the material, the greater the absorption of radiation. If the paper is too thick, less radiation passes through to the detector and the rollers move closer together.

**Treating Cancer**

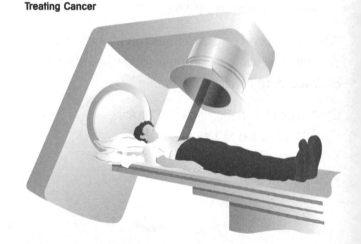

**Controlling Thickness**

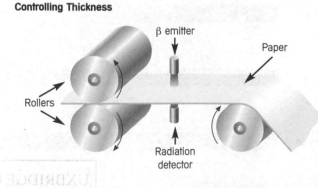

## Key Words

**Half-life • Ionising power • Radiation**

## Acute Radiation Dangers

The **ionising power** of alpha, beta and gamma radiation can damage molecules inside healthy living cells. This results in the death of the cell.

Damage to cells in organs can cause **cancer**. The larger the dose of radiation, the greater the risk of cancer.

The damaging effect of radiation depends on whether the source is inside or outside the body.

If the source is **inside the body**…

- α causes most damage as it is easily absorbed by cells, causing the most ionisation
- β and γ cause less damage as they are less likely to be absorbed by cells.

If the source is **outside the body**…

- α cannot penetrate the body; it is stopped by the skin
- β and γ can penetrate the body to reach the cells of organs, where they are absorbed.

**Inside the Body**

α
β
γ

**Outside the Body**

α    β    γ

## Half-Life

The **half-life** of a radioactive isotope is a measurement of the **rate of radioactive decay**.

Half-life is the time it takes for the number of nuclei in a sample of the isotope to halve.

Half-life can also be found by measuring the time it takes for the **count rate** of a sample containing the isotope to halve. (Count rate is the number of atoms which decay in a certain time.)

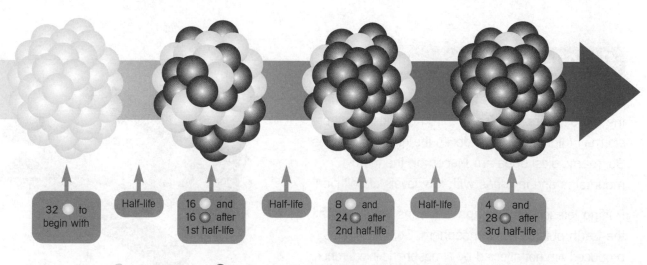

32 ○ to begin with
Half-life
16 ○ and 16 ● after 1st half-life
Half-life
8 ○ and 24 ● after 2nd half-life
Half-life
4 ○ and 28 ● after 3rd half-life

○ = original nuclei    ● = new nuclei formed after original nuclei have decayed

# The Universe

## Observing the Universe

Observations of the **solar system** and the **galaxies** in the **Universe** can be carried out…

* on Earth
* from space.

One method of observation is to use a **telescope**. Different types of telescope can detect **visible light** or other **electromagnetic radiations**, e.g. radio waves or X-rays.

## Telescopes

A **reflecting telescope reflects** light using **mirrors**. It can only be used at night.

A **refracting telescope refracts** light at each end using **lenses**. It produces sharp, detailed images but can only be used at night.

**Radio telescopes** pick up **radio waves** instead of light waves.

Radio waves are much longer than light waves, so in order to receive good signals, radio telescopes need large **antennae** or lots of smaller antennae working together.

The radio waves are emitted by bodies in space.

Most radio telescopes use a **parabolic dish** to reflect the radio waves to a **receiver**, which detects and amplifies the signal.

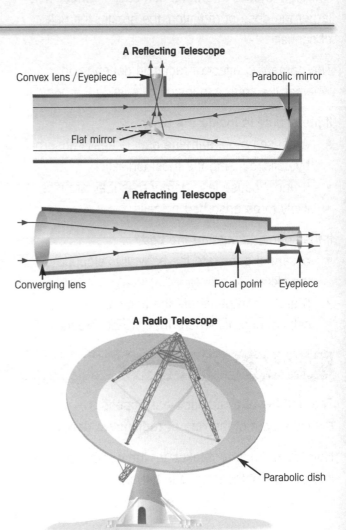

**A Reflecting Telescope**

Convex lens / Eyepiece — Parabolic mirror — Flat mirror

**A Refracting Telescope**

Converging lens — Focal point — Eyepiece

**A Radio Telescope**

Parabolic dish

## Using Telescopes

Observing the Universe **from the Earth** is limited by the **atmosphere**. Interference from clouds, weather storms or light pollution reduces the quality of images. So, many telescopes are placed on the top of mountains and in areas with low levels of pollution.

Putting telescopes **into space** means that they orbit the Earth outside the atmosphere. So, the images produced are not affected by atmospheric interference.

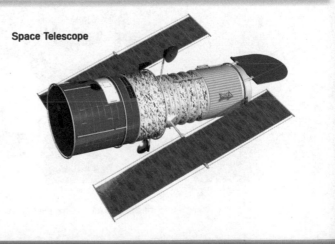

**Space Telescope**

## Red Shift

If a wave source is moving away from, or towards, an observer, there will be a change in the…

- observed **wavelength**
- observed **frequency**.

If a light source moves away from us, the wavelengths of the light in its spectrum are longer than if it wasn't moving.

This is known as **red shift**; the wavelengths 'shift' towards the red end of the **electromagnetic spectrum**.

There is a red shift in light observed from most distant galaxies.

This means that they're moving away from us very quickly.

This effect is exaggerated in galaxies which are further away. This means that the further away a galaxy is, the faster it's moving away from us.

This observed red shift suggests that…

- the whole Universe is **expanding**
- the Universe might have started billions of years ago, from one small place, with a huge explosion known as the '**Big Bang**'.

## Key Words

**Electromagnetic spectrum • Red shift • Reflection • Refraction • Telescope**

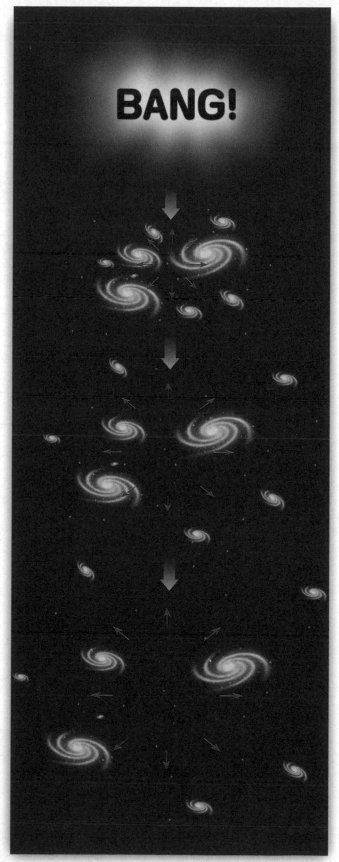

BANG!

# Unit 1b Summary

## Electromagnetic Radiation

Electromagnetic radiation travels as waves which form the **electromagnetic spectrum**:

Gamma rays | X-rays | Ultraviolet rays | **Visible light** | Infra red rays | Microwaves | Radio waves

High frequency, short wavelength

Low frequency, long wavelength

**Visible spectrum** = the only part of the electromagnetic spectrum that can be seen.

| Electromagnetic Wave | Uses | Effects |
|---|---|---|
| Radio waves | Transmitting signals. | High levels of exposure = tissue damage. |
| Microwaves | Satellite communication; cooking. | May damage or kill cells. |
| Infra red rays | Grills; toasters; remote controls; optical fibre communication. | Excess can cause burns. |
| Ultraviolet rays | Security coding; sunbeds. | Can kill cells or cause cancer. |
| X-rays | Producing shadow images of bones / metals; treating cancer. | Can kill cells or cause cancer. |
| Gamma rays | Treating cancer; sterilisation. | Can kill cells or cause cancer. |

## Analogue and Digital Signals

Analogue signals vary in amplitude and / or frequency.

Digital signals don't vary. They have two values: on (1) and off (0).

Digital signals are of a better quality than analogue.

## Isotopes

Isotopes = atoms of a particular element which have the same number of protons but different numbers of neutrons.

## Radiation

There are 3 types of radiation:

| Type | Description | Ionising Power |
|---|---|---|
| **Alpha** ($\alpha$) | Helium nucleus | Strong |
| **Beta** ($\beta$) | High-energy electron | Reasonable |
| **Gamma** ($\gamma$) | High-frequency electromagnetic radiation | Weak |

# Unit 1b Summary

Alpha and beta radiations are deflected by electric fields and magnetic fields.

Radiation can be used...
- for sterilising instruments
- for treating cancer
- as tracers
- for controlling the thickness of materials.

## Radiation Dangers

| Radiation | Inside Body | Outside Body |
| --- | --- | --- |
| Alpha (α) | Easily absorbed by cells. | Cannot penetrate skin. |
| Beta (β) | Less likely to be absorbed. | Penetrates skin and is absorbed by cells of organs. |
| Gamma (γ) | Less likely to be absorbed. | Penetrates skin and is absorbed by cells of organs. |

## Half-Life

Half-life = the time it takes for the number of nuclei of the radioactive isotope in a sample to halve.

## Telescopes

Observations of space can be carried out using telescopes.

**Refracting telescope** refracts light using **lenses**.
**Reflecting telescope** reflects light using **mirrors**.
**Radio telescope** picks up **radio waves** using a parabolic dish.

Observing the Universe from Earth is limited by atmospheric interference from clouds, storms, pollution, etc.

Observing the Universe from space is not affected by atmospheric interference.

## Red Shift

Red shift = wavelengths of light 'shift' towards the red end of the electromagnetic spectrum.

Red shift suggests that...
- the Universe is expanding
- the Universe began with a **Big Bang**.

# Unit 1b Practice Questions

**1** The electromagnetic spectrum can be arranged in order of wavelength.

Match the types of electromagnetic radiation A, B, C and D with the numbers 1 to 4 in the diagram of the electromagnetic spectrum below.

**A** X-rays ........................

**B** microwaves ........................

**C** visible light ........................

**D** gamma rays ........................

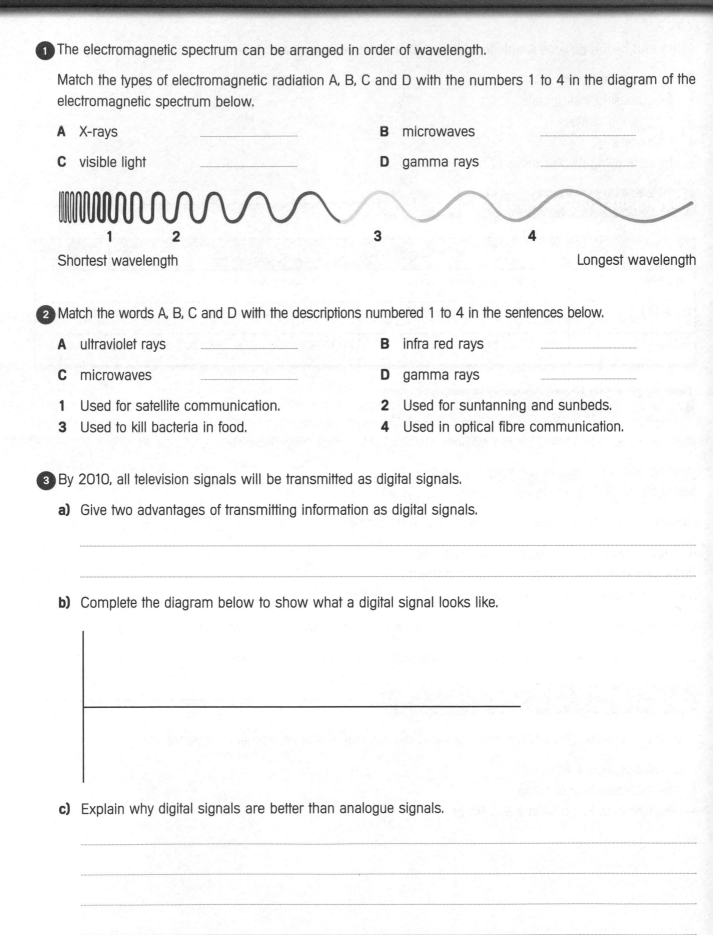

1          2                           3                    4

Shortest wavelength                                                    Longest wavelength

**2** Match the words A, B, C and D with the descriptions numbered 1 to 4 in the sentences below.

**A** ultraviolet rays ........................

**B** infra red rays ........................

**C** microwaves ........................

**D** gamma rays ........................

1   Used for satellite communication.
3   Used to kill bacteria in food.

2   Used for suntanning and sunbeds.
4   Used in optical fibre communication.

**3** By 2010, all television signals will be transmitted as digital signals.

**a)** Give two advantages of transmitting information as digital signals.

........................................................................................................................

........................................................................................................................

**b)** Complete the diagram below to show what a digital signal looks like.

**c)** Explain why digital signals are better than analogue signals.

........................................................................................................................

........................................................................................................................

........................................................................................................................

........................................................................................................................

**4** Match the words A, B, C and D with the spaces numbered 1 to 4 in the sentences below.

**A** beta ..................................... **B** radioactive .....................................

**C** frequency ..................................... **D** helium .....................................

A substance is __1__ when the structure of the atom changes and it gives out radiation. A radioactive substance that emits __2__ nuclei is called alpha radiation, whereas when a high-energy electron is emitted we get __3__ radiation. Gamma radiation is a type of high __4__ electromagnetic radiation.

**5** Different types of radiation have different penetrating powers.

Match the words A, B, C and D with the spaces numbered 1 to 4 in the sentences below.

**A** beta ..................................... **B** lead .....................................

**C** many ..................................... **D** paper .....................................

All radioactive substances are absorbed by __1__. __2__ radiation is able to penetrate paper but is stopped by a few millimetres of aluminium. Gamma radiation is the most penetrating and is only stopped by __3__ centimetres of lead. Alpha radiation in stopped by a thin sheet of __4__.

**6 a)** Explain how a beta source can be used on a machine controlling the thickness of wallpaper.

.................................................................................................................................................

.................................................................................................................................................

.................................................................................................................................................

**b)** Suggest two other uses of radiation.

.................................................................................................................................................

**7** What is meant by 'half-life of a radioactive substance'?

.................................................................................................................................................

.................................................................................................................................................

.................................................................................................................................................

**8** Match the words A, B, C and D with the descriptions numbered 1 to 4 below.

**A** refracting telescope ..................... **B** reflecting telescope .....................

**C** radio telescope ..................... **D** telescope in space .....................

**1** Uses lenses to refract light. **2** Uses a parabolic mirror.
**3** Uses a parabolic dish. **4** Isn't affected by atmospheric interference.

# Speed

## Speed

The **speed** of an object is a measure of **how fast** it is moving.

To work out the speed of any moving object, you need to know…

- the **distance** it travels
- the **time it takes** to travel that distance.

The cyclist below travels a **distance** of 8 metres every second. So, you can say that his speed is 8 metres per second (m/s).

You can calculate speed using this formula…

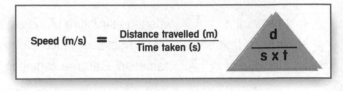

$$\text{Speed (m/s)} = \frac{\text{Distance travelled (m)}}{\text{Time taken (s)}}$$

Speed can be measured in…

- metres per second (m/s)
- kilometres per hour (km/h)
- miles per hour (mph).

## Distance–Time Graphs

The **slope** of a **distance–time graph** represents the **speed** of an object. The **steeper the slope**, the **greater the speed**.

The graph shows…

1. a stationary person
2. a person moving at a constant speed of 2m/s
3. a person moving at a greater constant speed of 3m/s.

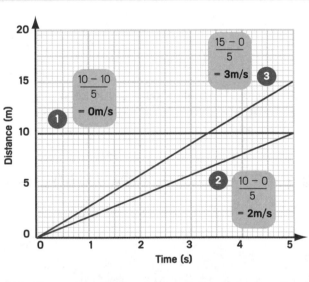

$$\text{①} \quad \frac{10 - 10}{5} = 0\text{m/s}$$

$$\text{③} \quad \frac{15 - 0}{5} = 3\text{m/s}$$

$$\text{②} \quad \frac{10 - 0}{5} = 2\text{m/s}$$

*N.B. The vertical axis shows distance from a fixed point (O), not total distance travelled.*

## Velocity

**Velocity** and speed **aren't** the same thing. The velocity of a moving object describes **its speed** in a **given direction**.

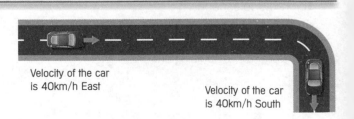

Velocity of the car is 40km/h East

Velocity of the car is 40km/h South

## Acceleration

The **acceleration** of an object is the **rate** at which its **velocity changes**. It is a measure of how quickly an object **speeds up** or **slows down**.

To work out the acceleration of any moving object, you need to know...

- the **change in velocity**
- the **time taken** for this change in velocity.

The cyclist below increases his velocity by 2 metres per second every second. So, his acceleration is 2m/s².

- **His velocity increases** by the **same amount** every second.
- The **actual distance** travelled each **second increases**.

You can calculate acceleration by using this formula:

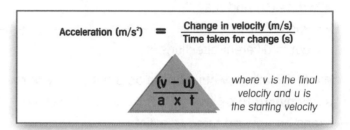

Acceleration (m/s²) $= \dfrac{\text{Change in velocity (m/s)}}{\text{Time taken for change (s)}}$

$\dfrac{(v - u)}{a \times t}$

*where v is the final velocity and u is the starting velocity*

**Deceleration** is a negative acceleration. It describes an object that is **slowing down**. It is calculated using the same formula.

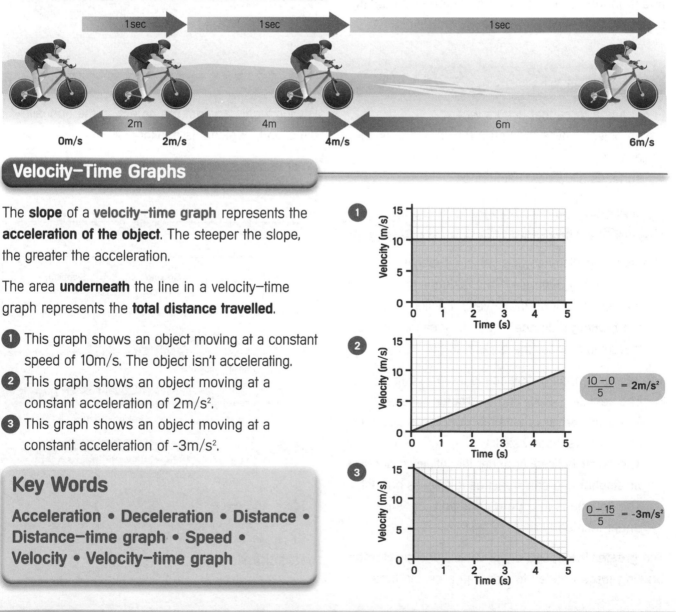

1sec · 1sec · 1sec

2m · 4m · 6m

0m/s · 2m/s · 4m/s · 6m/s

## Velocity–Time Graphs

The **slope** of a **velocity–time graph** represents the **acceleration of the object**. The steeper the slope, the greater the acceleration.

The area **underneath** the line in a velocity–time graph represents the **total distance travelled**.

❶ This graph shows an object moving at a constant speed of 10m/s. The object isn't accelerating.

❷ This graph shows an object moving at a constant acceleration of 2m/s².

❸ This graph shows an object moving at a constant acceleration of -3m/s².

**①** Velocity (m/s) vs Time (s)

**②** Velocity (m/s) vs Time (s) — $\dfrac{10 - 0}{5} = 2m/s^2$

**③** Velocity (m/s) vs Time (s) — $\dfrac{0 - 15}{5} = -3m/s^2$

## Key Words

**Acceleration • Deceleration • Distance • Distance–time graph • Speed • Velocity • Velocity–time graph**

# Forces

## Forces

**Forces** are pushes or pulls. They are measured in **newtons** (N) and may…

- vary in size
- act in different directions.

When a stationary object rests on a surface it exerts a **downward force** (weight). The surface it rests on exerts an **upward force** (reaction).

These two forces are **equal and opposite** and the object remains stationary.

A number of forces acting on an object can be replaced by a **single force**. This force has the **same effect** on the object as the **original forces** all **acting together**. This is called the **resultant force**.

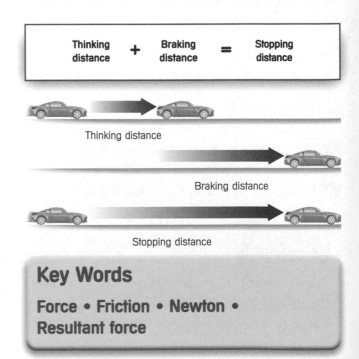

Upward force (reaction)

Downward force (weight)

## Friction

**Friction** is a force that occurs when…

- an object moves through a medium, e.g. air or water
- surfaces slide past each other.

Friction works **against the object**, in the opposite direction to which it is moving.

When a motor vehicle travels at a steady speed, the **frictional forces balance the driving force**.

## Stopping Distance

The stopping distance of a vehicle depends on…

- the **thinking distance** – the distance travelled by the vehicle during the driver's reaction time
- the **braking distance** – the distance travelled by the vehicle under the braking force.

The overall stopping distance is increased if:

- The vehicle is travelling at **greater speeds**.
- There are **adverse weather conditions**, e.g. wet or icy roads, poor visibility.
- The **driver is tired**, or under the influence of **drugs or alcohol** and can't react as quickly as normal.
- The **vehicle** is in a **poor condition**, e.g. it has under-inflated tyres.

The **greater the speed** of the vehicle, the **greater the braking force** needed to stop it in a certain time.

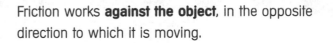

| Thinking distance | + | Braking distance | = | Stopping distance |

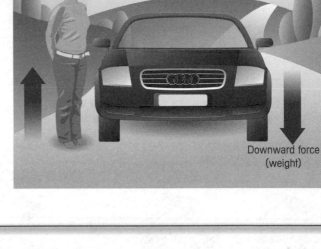

Thinking distance

Braking distance

Stopping distance

## Key Words

Force • Friction • Newton • Resultant force

## How Forces Affect Movement

The **movement** of an object depends on the **forces** acting on it.

If the forces are **equal and opposite** then they are **balanced**, i.e. the resultant force is **zero**.

If the forces are **not equal and opposite** then they are **unbalanced**, i.e. the resultant force is **not zero**.

If the resultant force acting on a **stationary** object is…

1. **zero**, the object will remain stationary

2. **not zero**, the object will start to move in the direction of the resultant force.

If the resultant force acting on a **moving object** is…

3. **zero**, the object will continue at the same speed in the same direction

4. **not zero**, the object will speed up or slow down in the direction of the resultant force.

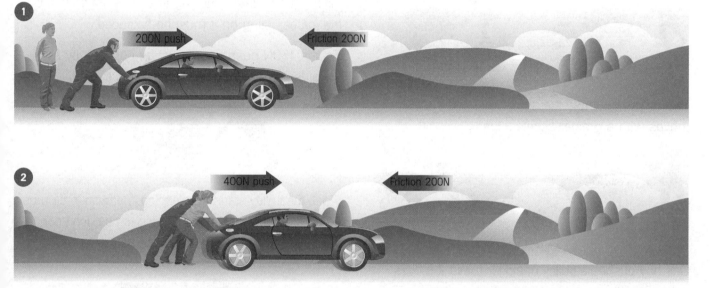

1. 200N push — Friction 200N

2. 400N push — Friction 200N

3. 200N push — Friction 200N

4. ON push — Friction 200N

# Force, Mass and Acceleration

## Force, Mass and Acceleration

If an unbalanced force acts on an object, the **acceleration** of the object will depend on…

- the **size** of the unbalanced force – the bigger the force, the greater the acceleration
- the **mass** of the object – the bigger the mass, the smaller the acceleration.

### Example

One boy pushes a trolley. He exerts an unbalanced force which causes the trolley to move and accelerate.

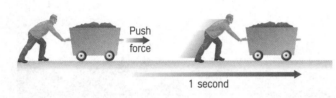

1 second

Two boys now push the same trolley. The trolley moves with a greater acceleration.

1 second

One boy pushes a trolley of bigger mass. It moves with a smaller acceleration than the first trolley.

1 second

The relationship between force, mass and acceleration is shown in this formula:

| Resultant force (N) | = | Mass (kg) | X | Acceleration (m/s²) | F / (m x a) |
|---|---|---|---|---|---|

So, a **newton** (N) is the force needed to give a mass of one kilogram an acceleration of one metre per second squared (1m/s²).

### Example

A trolley of mass 400kg is pushed along a floor with a constant speed by one boy who exerts a push force of 150N.

Another boy joins him to increase the push force, and the trolley accelerates at 0.5m/s².

Calculate…

**a)** the force needed to achieve this acceleration
**b)** the total push force exerted on the trolley.

Initially the trolley is moving at a constant speed, so the forces acting on it must be balanced.

So, the 150N push force must be opposed by an equal force, i.e. friction or air resistance.

Constant speed — 150N

When the trolley starts accelerating, the push force must be greater than friction, etc. These forces don't cancel each other out and an unbalanced force now acts.

Push force — Acceleration of 0.5m/s² — Friction 150N

**a)** Force = Mass x Acceleration
= 400kg x 0.5m/s²
= **200N**

**b)** Total push = Force needed to equal friction + Force needed to provide acceleration
= 150N + 200N
= **350N**

## Key Words

**Acceleration • Mass • Newton • Resistance • Resultant force • Terminal velocity • Weight**

## Terminal Velocity

Falling objects experience two forces…
- the **downward force of weight**, W (↓)
- the **upward force of air resistance**, R, or drag (↑).

Weight always stays the same, but the force of air resistance changes.

**Gravity** is a force of attraction that acts between objects that have mass, e.g. a falling object and the Earth. The weight of an object is the force exerted on it by gravity. It is measured in **newtons** (N).

If a skydiver, Nick, jumps out of an aeroplane, the speed of his descent can be considered in two separate parts:
- before the parachute opens
- after the parachute opens.

### Before the Parachute Opens

**1** Immediately after Nick jumps, he accelerates due to the force of **gravity**.

**2** As he falls, he experiences the **frictional force of air resistance** (R) in the opposite direction. At this point **weight** (W) is greater than R, so he continues to accelerate.

**3** As his speed increases, so does R.

**4** R increases until it is equal to W. The **resultant force** acting on Nick is now zero and his falling speed becomes **constant**. This speed is called the **terminal velocity**.

### After the Parachute Opens

**5** When the parachute opens, the force of R is now greatly increased and is bigger than W.

**6** The increase in R decreases Nick's speed. As his speed decreases, so does R.

**7** R decreases until it is equal to W. The forces acting are once again balanced and, for the second time, he falls at a steady speed, although slower than before. This is a **new terminal velocity**.

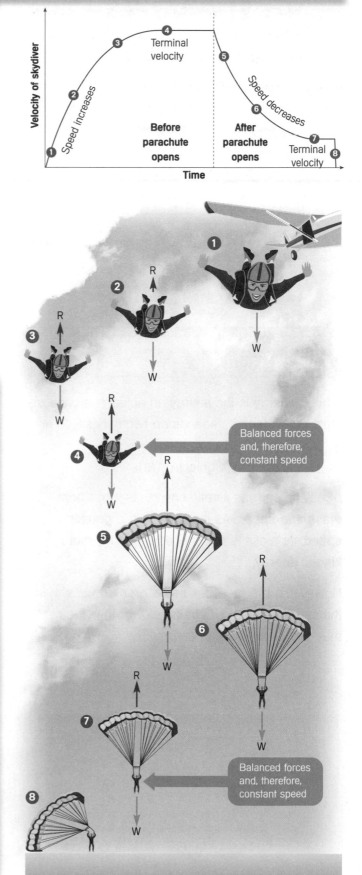

# Work and Kinetic Energy

## Work

When a force moves an object, **work** is **done** on the object resulting in the **transfer** of **energy**.

Energy is measured in joules (J). So…

> Work done (J) **=** Energy transferred (J)

The relationship between work done, force and distance is shown by this formula:

> Work done (J) **=** Force applied (N) **X** Distance moved in direction of force (m)
>
> $$\frac{W}{F \times d}$$

### Example

A man pushes a car with a steady force of 250N. The car moves a distance of 20m. How much work does the man do?

Work done = Force applied x Distance moved
= 250N x 20m
= **5000J (or 5kJ)**

Work done against frictional forces is mainly transformed into **heat energy**. When work is done on an elastic object to change its shape, the energy is stored in the object as **elastic potential energy**.

## Kinetic Energy

**Kinetic energy** is the energy an object has because of its **movement**. It depends on two things:
- the **mass** of the object (kg)
- the **speed** of the object (m/s).

A moving car has kinetic energy as it has **both mass and speed**. As it moves with a **greater speed**, its mass is unchanged but it has **more kinetic energy**.

However, a moving truck has a **greater mass** than the car. So, even if it is going slower than the car, it may have **more kinetic energy**.

(HT) You can calculate kinetic energy by using this formula:

> Kinetic energy (J) **=** $\frac{1}{2}$ **X** Mass (kg) **X** Speed$^2$ (m/s)$^2$
>
> $$\frac{K.E.}{\frac{1}{2} \times m \times v^2}$$

### Example

A car of mass 1000kg is moving at a speed of 10m/s. How much kinetic energy does it have?

Kinetic energy = $\frac{1}{2}$ x Mass x Speed$^2$

= $\frac{1}{2}$ x 1000kg x (10m/s)$^2$

= **50 000J (or 50kJ)**

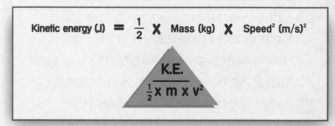

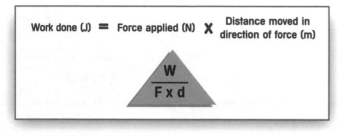

## Momentum

Momentum is a measure of the state of **motion** of an object. It depends on two things:
- the **mass** of the object (kg)
- the **velocity** of the object (m/s).

A moving car has momentum as it has **both mass and velocity**. If the car moves with a **greater velocity**, it will have **more momentum** providing its mass hasn't changed.

However, a moving truck with a greater mass may have more momentum than the car even if its velocity is less.

You can calculate momentum by using this formula:

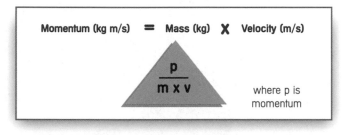

Momentum (kg m/s) = Mass (kg) X Velocity (m/s)

$$\frac{p}{m \times v}$$

where p is momentum

### Example 1

Calculate the momentum of a car of mass 1200kg that is travelling at a velocity of 30m/s.

Momentum = Mass x Velocity
= 1200kg x 30m/s
= **36 000kg m/s**

### Example 2

A truck has a mass of 4000kg. Calculate its velocity if it has the same momentum as the car in Example 1.

Rearrange the formula…

$$\text{Velocity} = \frac{\text{Momentum}}{\text{Mass}}$$

$$= \frac{36\,000\text{kg m/s}}{4000\text{kg}}$$

= **9m/s**

N.B. Since the truck has a **greater mass** than the car, it can move at a **slower speed** and still have the **same momentum**.

## Magnitude and Direction

Momentum has…
- **magnitude** (size)
- **direction**.

The direction of movement is especially important when calculating momentum. For example…
- car A (moving from left to right) has a positive velocity and, consequently, a positive momentum
- car B (moving from right to left) has a negative velocity and negative momentum because it is moving in the opposite direction to car A.

### Examples

Car A has a mass of 1000kg and a velocity of 20m/s. Calculate its momentum.

Momentum = 1000kg x 20m/s
= **20 000kg m/s**

Car B has a mass of 1000kg and a velocity of 20m/s. Calculate its momentum.

Momentum = 1000kg x -20m/s
= **-20 000kg m/s**

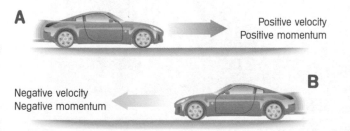

A
Positive velocity
Positive momentum

Negative velocity
Negative momentum
B

### Key Words

Energy • Kinetic energy • Mass • Momentum • Speed • Transfer • Velocity • Work

# Momentum

## Force and Change in Momentum

When a **force** acts on a **moving object**, or a stationary object that is capable of moving, the object will experience a **change** in **momentum**.

An external force can…

- give a stationary object **momentum**
- **increase** the momentum of a moving object
- **decrease**, or stop, the momentum of a moving object.

The extent of the change in momentum depends on…

- the **size** of the force
- the **length** of time the force is acting on the object.

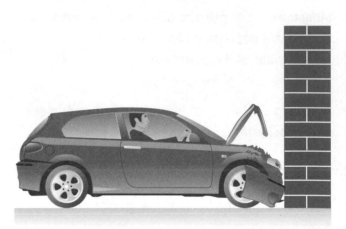

**(HT)** Force, change in momentum and the time taken for the change are related by this formula:

Force (N) = Change in momentum (kg m/s) / Time taken for change (s)

where Δ(mv) is change in momentum

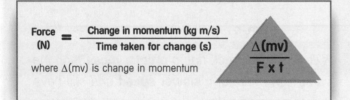

$$\frac{\Delta(mv)}{F \times t}$$

### Example
A girl kicks a stationary ball with a force of 30N. The force acts on the ball for 0.15 seconds. The mass of the ball is 0.5kg.

**a)** Calculate the change in momentum of the ball

Change in momentum = Force x Time

                = 30N x 0.15s

                = **4.5kg m/s**

**b)** Calculate the increase in velocity of the ball.

Velocity = $\frac{\text{Momentum}}{\text{Mass}}$ = $\frac{4.5\text{kg m/s}}{0.5\text{kg}}$ = **9m/s**

The girl can **increase** the **change in momentum** of the ball, and as a result its **velocity**, without increasing the force applied. She can do this by 'following through' with her kick. This will **increase the time** for which the **force is applied**.

Some sports where 'following through' increases the velocity of the ball are…

- football
- cricket
- golf
- tennis
- squash.

# Collisions and Explosions

## Collisions and Explosions

In any collision or explosion, the **momentum** in a particular direction **after the event** is the same as the momentum in that direction **before the event**.

**Momentum is conserved**, provided that no external forces act.

### Example 1

Two cars are travelling in the same direction along a road. Car A collides with the back of car B and they stick together. Calculate their velocity after the collision.

**Before**

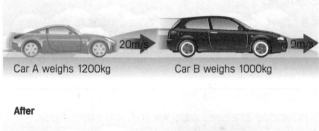

Car A weighs 1200kg          Car B weighs 1000kg

**After**

Car A + Car B weighs 2200kg

**Momentum before collision:**

= Momentum of A + Momentum of B

= (Mass x velocity of A) + (Mass x velocity of B)

= (1200kg x 20m/s) + (1000kg x 9m/s)

= 24 000kg m/s + 9000kg m/s

= 33 000kg m/s

**Momentum after collision:**

= Momentum of A and B

= (Mass of A + Mass of B) x (Velocity of A + B)

= (1200+1000) x v

= 2200v

**Since momentum is conserved:**

Total momentum before = Total momentum after

$$33\,000 = 2200v$$

$$\text{So, } v = \frac{33\,000}{2200}$$

$$= \textbf{15m/s}$$

### Example 2

A gun fires a bullet of mass 0.01kg. The velocity of the bullet is 350m/s. Calculate the recoil velocity of the gun.

**Before**          **After**

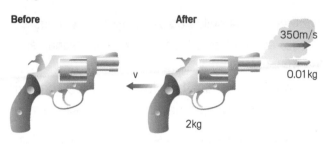

350m/s

0.01kg

v

2kg

Firing a gun is an example of an explosion where two objects move **away** from each other.

Since the gun and the bullet are moving in opposite directions, we will assume that the **bullet** has **positive** velocity and momentum which means that the **gun** has **negative** velocity and momentum.

**Momentum before explosion = 0**

(Neither the gun nor the bullet are moving.)

**Momentum after explosion:**

= Momentum of bullet + Momentum of gun

= (Mass x velocity of bullet) + (Mass x velocity of gun)

= (0.01kg x 350m/s) + (2kg x -v)

= 3.5 − 2v

> Remember the gun has negative velocity

**Since momentum is conserved:**

Momentum after explosion = Momentum before explosion

Momentum of bullet + Momentum of gun = 0

(Mass x velocity of bullet) + (Mass x velocity of gun) = 0

(0.01kg x 350m/s) + (2kg x v) = 0

$$3.5 + 2v = 0$$

$$2v = -3.5$$

$$v = \frac{3.5}{2}$$

$$v = \textbf{-1.75m/s}$$

## Key Words

**Force • Momentum**

# Static Electricity

## Static Electricity

Some insulating materials can become **electrically charged** when they are rubbed against each other. Unless it is **discharged**, the electrical charge (static) stays on the material.

**Static electricity** builds up when **electrons** (negative charge) are 'rubbed off' one material onto another. The material…

- **receiving** electrons becomes **negatively** charged
- **giving up** electrons becomes **positively** charged.

**Examples**

A perspex rod rubbed with cloth loses electrons to become positively charged. The cloth gains electrons to become negatively charged.

An ebonite rod rubbed with fur gains electrons to become negatively charged. The fur loses electrons to become positively charged.

## Repulsion and Attraction

When two charged materials are brought together, they **exert a force** on each other. They are either **attracted** or **repelled**:

- Materials with the **same** charge **repel** each other.
- Materials with **different** charge **attract** each other.

**Examples**

If a charged perspex rod is moved near to a suspended perspex rod, the suspended perspex rod will be repelled.

If a charged ebonite rod is moved near to a suspended charged perspex rod, the suspended perspex rod will be attracted.

## The Uses of Static

**Electrostatic** charges can be very useful in industry and at home. They can be used in a variety of different ways.

**Examples**

Reducing air pollution in **electrostatic smoke precipitators**, and reproducing images in **photocopiers**.

## Electrostatic Smoke Precipitator

Smoke precipitators are designed to remove solid smoke particles from waste gases before the gases are released into the environment:

1. Solid smoke particles become positively charged as they pass by a charged metal grid.
2. These 'like' charges repel each other causing the particles to move away from the grid.
3. The particles are then attracted to the negatively charged collecting plates. The particles stick to form a layer of 'soot' which is regularly removed.

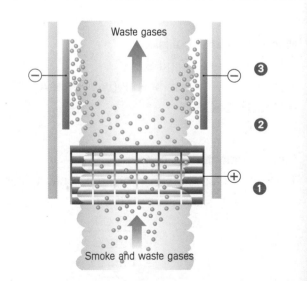

Waste gases

Smoke and waste gases

## The Photocopier

1. An image of the page to be copied is projected onto an electrically charged plate (usually positively charged).
2. Light causes charge to leak away leaving an electrostatic impression of the page.
3. The charged impression on the plate attracts tiny specks of black powder. The powder is then transferred from the plate to paper. Heat is used to fix the final image on the paper.

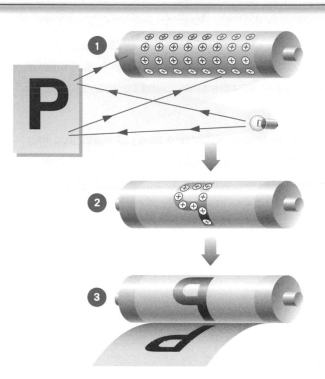

### Key Words

**Attraction • Charged • Current • Electron • Electrostatic • Repulsion • Static electricity**

## Discharge of Static Electricity

A **charged conductor** (positive or negative) can be **discharged**. Any charge on it is **removed** by connecting it to earth with a conductor.

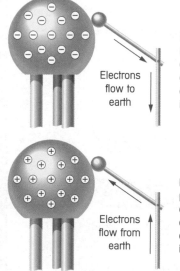

Electrons flow to earth

If a conductor touches a negatively charged dome, electrons flow from the dome to earth, via the conductor, until the dome is completely discharged.

Electrons flow from earth

If a conductor touches a positively charged dome, electrons flow from earth to cancel out the positive charge on the dome, until the dome is completely discharged.

This flow of electrons through a solid conductor is an **electric current**.

Metals **conduct electricity** well because electrons from their **atoms** can **move freely** throughout the metal structure.

**HT** The greater the charge on an isolated object, the greater the **potential difference** between the object and earth.

If the potential difference between the object and a nearby earthed conductor becomes high enough, the air molecules can **ionise**, and there is a spark as **discharge** occurs.

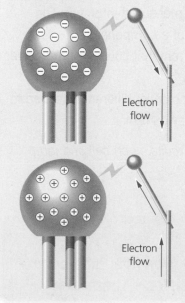

Electron flow

If an earthed conductor is near a dome with a high negative charge, electrons flow from the dome through the air to get to earth via the conductor.

Electron flow

If an earthed conductor is near a dome with a high positive charge, electrons flow from the dome through the air to cancel out the positive charge on the dome.

# Electric Circuits

## Circuits

An electric **current** will flow through an electrical component (or device) if there is a **potential difference** (voltage) across the ends of the component.

In a circuit...

- the potential difference (p.d.) is measured in **volts (V)** using a **voltmeter** connected in **parallel**
- the **current** is measured in amperes (A), using an ammeter connected in series.

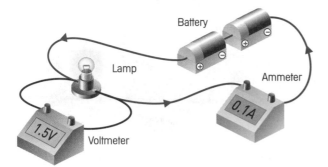

The amount of current that flows through the component depends on...

- the potential difference across the component
- the **resistance** of the component.

The **greater the potential difference** across a component, the **greater the current** that flows through the component.

Components resist the flow of current through them. They have resistance (measured in ohms). The greater the resistance of a component(s)...

- the smaller the current that flows for a particular potential difference
- **or** the greater the potential difference needed to maintain a particular current.

*N.B. In the following circuits, each cell and lamp are identical.*

### Key Words

**Current • Diode • Potential difference • Resistance • Resistor • Thermistor**

### Circuit 1

The cell provides p.d. across the lamp. A current flows and the lamp lights up.

### Circuit 2

Two cells together provide a bigger p.d. across the lamp. A bigger current flows and the lamp lights up more brightly (compared to circuit 1).

### Circuit 3

Two lamps together have a greater resistance. A smaller current now flows and the lamps light up less brightly (compared to circuit 1).

### Circuit 4

Two cells together provide a greater p.d. The same current as in circuit 1 now flows. The lamps light up as brightly as in circuit 1.

## Resistance

Resistance is a measure of **how hard** it is to **get a current** through a **component** at a particular potential difference.

**Current–potential difference graphs** show how the current through the component varies with the potential difference across it.

Potential difference, current and resistance are related by this formula:

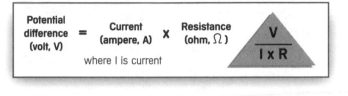

| Potential difference (volt, V) | = | Current (ampere, A) | X | Resistance (ohm, Ω) |

where I is current

$$\frac{V}{I \times R}$$

## Resistors

| | |
|---|---|
| The resistance of a **light dependent resistor (LDR)** depends on the amount of light falling on it. Its **resistance decreases** as the amount of **light** falling on it **increases**. This allows more current to flow. | |
| The resistance of a **thermistor** depends on its **temperature**. Its **resistance decreases** as the **temperature** of the thermistor **increases**. This allows more current to flow. | |
| As long as the temperature of the **resistor** stays constant, the current through the resistor is directly proportional to the voltage across the resistor. This is regardless of which direction the current is flowing, i.e. if one doubles, the other also doubles. | |
| As the temperature of the **filament lamp** increases, and the bulb gets brighter, then the resistance of the lamp increases. This is regardless of which direction the current is flowing. | |
| A **diode** allows a current to flow through it in **one direction only**. It has a very high resistance in the reverse direction so no current flows. | |

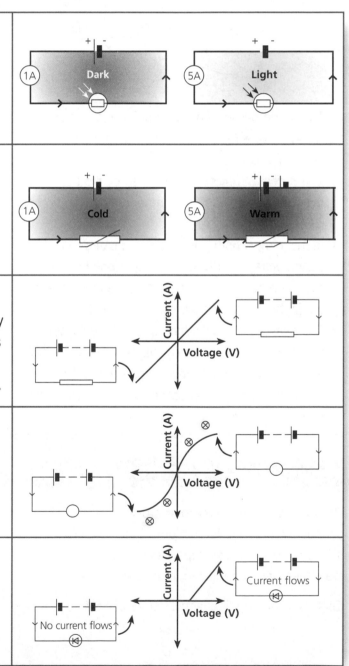

# Electric Circuits

## Connecting Components in Series

In a **series circuit**, all components are connected one after the other in one loop, going from one terminal of the battery to the other.

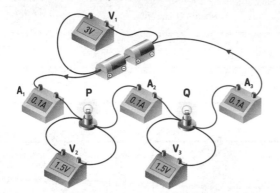

- The **same current** flows through each component, i.e. $A_1 = A_2 = A_3$. In this circuit, each ammeter reading is 0.1A.
- The **potential difference** supplied by the battery is **divided** up between the components in the circuit, i.e. $V_1 = V_2 + V_3$.
- In this circuit, both bulbs have the same resistance so the voltage is divided equally. If one bulb had twice the resistance of the other, then the voltage would be divided differently, e.g. 2V and 1V.
- The total resistance is the sum of the individual resistances of the components, i.e. $\Omega = \Omega_p + \Omega_Q$. If both P and Q each have a resistance of 15 ohms, the total resistance would be 15 ohms + 15 ohms = 30 ohms.

## Connecting Components in Parallel

In a **parallel circuit**, all components are connected separately in their own loop going from one terminal of the battery to the other.

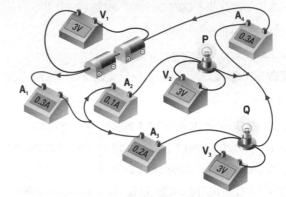

- The total current in the main circuit is equal to the sum of the currents through the separate components, i.e. $A_1 = A_2 + A_3 = A_4$. In this circuit, 0.3A = 0.1A + 0.2A = 0.3A.
- The potential difference across each component is the same (and is equal to the p.d. of the battery), i.e. $V_1 = V_2 = V_3$. In this circuit, each bulb has a p.d. of 3V across it.
- The amount of current which passes through a component depends on the resistance of the component. The greater the resistance, the smaller the current. Bulb P must have twice the resistance of bulb Q as only 0.1A passes through bulb P while 0.2A passes through bulb Q.

## Connecting Cells in Series

The **total potential difference** provided by cells connected in series is the sum of the potential difference of each cell separately, providing that they have been connected in the same direction.

Each cell has a potential difference of 1.5V.

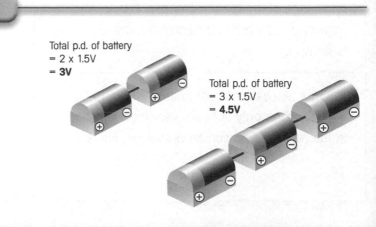

Total p.d. of battery
= 2 x 1.5V
= **3V**

Total p.d. of battery
= 3 x 1.5V
= **4.5V**

## Currents

A **direct current** (d.c.) always flows in the **same direction**. Cells and batteries supply direct current.

An **alternating current** (a.c.) **changes the direction** of flow back and forth continuously.

The **frequency** is the number of complete cycles of reversal per second. The UK mains electricity is 50 cycles per second (Hertz).

The UK mains supply has a voltage of about **230 volts**. This voltage, if it isn't used safely, can kill.

## The Three-Pin Plug

**The 3-pin Plug**

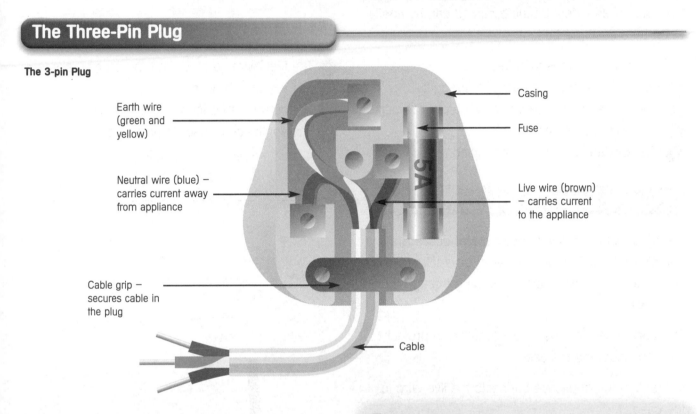

Earth wire (green and yellow)

Neutral wire (blue) – carries current away from appliance

Cable grip – secures cable in the plug

Casing

Fuse

5A

Live wire (brown) – carries current to the appliance

Cable

Most electrical appliances are connected to the mains electricity supply using a **cable** and a **three-pin plug**.

The plug is inserted into a socket on the ring main circuit.

The materials used for the plug and cable depend on their properties:
- The **inner cores** of the wires are made of **copper** because it's a **good conductor**.
- The **outer layers** are made of flexible **plastic** because it's a **good insulator**.
- The **pins** of a plug are made from **brass** because it's a **good conductor**.
- The **casing** is made from **plastic or rubber** because both are **good insulators**.

**HT** The live terminal of the mains supply alternates between a **positive** and **negative** voltage with respect to the neutral terminal.

The neutral terminal stays at a voltage close to zero with respect to earth.

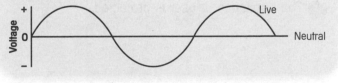

Voltage | + | 0 | − | Live | Neutral

## Key Words

**Alternating current • Direct current • Frequency • Parallel circuit • Series circuit**

# Circuit Breakers and Fuses

## Circuit Breakers and Fuses

If an electrical fault causes a **current that is too high**, the circuit will be **switched off** by a **circuit breaker** or **fuse**.

A **circuit breaker** is a safety device which **automatically breaks** an electric circuit if it becomes overloaded. Circuit breakers can be **easily reset** by pressing a button.

If the current becomes too high:
1. An electromagnet increases in strength.
2. The electromagnet separates a pair of contacts.
3. The circuit is broken ('trips').
4. The appliance or user is protected.

A **fuse** should always be part of a live circuit. It's a short, thin piece of wire with a **low melting point**.

The current rating of the fuse must be **just above** the normal working current of the appliance for the safety system to work properly.

If the current running through the appliance is greater than the current rating of the fuse:
1. The fuse wire gets hot and melts or breaks.
2. The circuit is broken.
3. The current is no longer able to flow.
4. The appliance or user is protected.

## Earthing

Electrical appliances with outer **metal cases** are usually **earthed**. The outer case of an appliance is connected to the earth pin in the plug through the earth wire.

The **earth wire** and **fuse** work together to protect the appliance and the user.

If a fault in the appliance connects the live wire to the case…
1. The case will become live.
2. The current will then 'run to earth' through the earth wire as this offers least resistance.
3. This overload of current will cause the fuse to melt (or the circuit breaker to trip).
4. The appliance or user is protected.

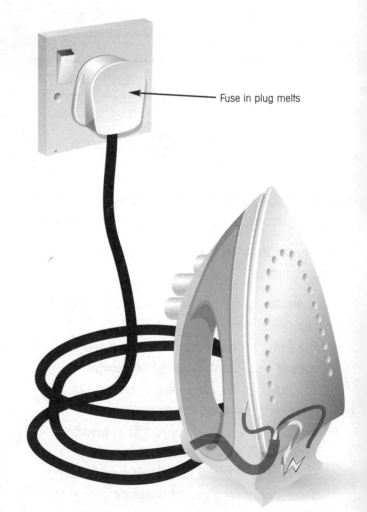

Fuse in plug melts

## Key Words

**Circuit breaker • Current • Earthed • Fuse • Resistor • Transform**

## Power

An electric **current** is the rate of **flow of charge**. A current transfers **electrical energy** from a battery or power supply to components in a **circuit**.

The rate of flow is measured in **amperes** (A).

The **components transform** some of this electrical energy into **other forms** of energy. For example, a **resistor** transforms **electrical** energy into **heat energy**.

The rate at which energy is transformed by a component or appliance is called the **power**.

You can calculate power by using this formula:

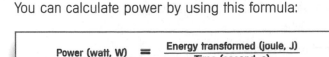

$$\text{Power (watt, W)} = \frac{\text{Energy transformed (joule, J)}}{\text{Time (second, s)}}$$

You can also calculate power using the formula:

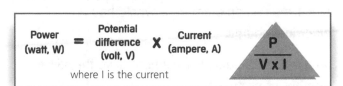

$$\text{Power (watt, W)} = \text{Potential difference (volt, V)} \times \text{Current (ampere, A)}$$

where I is the current

## HT Charge

The **amount of electrical charge** which passes any point in a circuit is measured in **coulombs** (C).

You can calculate charge using this formula:

$$\text{Charge (coulomb, C)} = \text{Current (amp, A)} \times \text{Time (second, s)}$$

where Q is charge

**Example:**

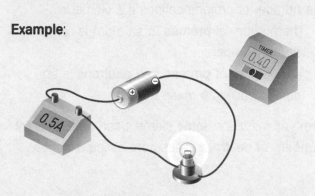

If the circuit above is switched on for 40 seconds and the current is 0.5 amps, what is the charge?

Charge = Current x Time
       = 0.5A x 40s
       **= 20 coulombs**

## HT Transforming Energy

As charge passes through a device, energy is **transformed**.

The **amount of energy** transformed for every coulomb of charge depends on the **size of the potential difference**. The greater the potential difference, the more energy transformed per coulomb.

**Example:**
If a circuit has a potential difference of 1.5V, and a charge of 23C, how much energy is transformed?

Energy transformed = p.d. x Charge
                 = 1.5V x 24C
                 **= 36 joules**

Remember, the charge gained this energy from the battery. It was transformed to the bulb whilst the circuit was switched on.

# Atoms and Radiation

## Atoms

An atom is made up of three parts:
- **protons**
- **neutrons**
- **electrons**.

An atom has the **same number** of **protons** as **electrons**. So, an atom as a whole has no **electrical charge**.

All atoms of a particular element have the **same number of protons**.

Atoms of different elements have **different numbers of protons**.

The number of protons defines the element:
- The number of **protons** in an atom is called its **atomic number**.
- The number of **protons** and **neutrons** in an atom is called its **mass number**.

Some atoms of the same element can have **different numbers of neutrons**. These are called **isotopes**.

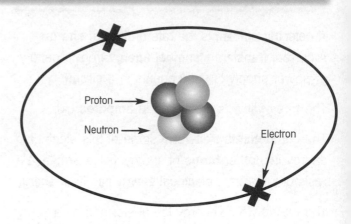

| Atomic Particle | Relative Mass | Relative Change |
|---|---|---|
| Proton | 1 | +1 |
| Neutron | 1 | 0 |
| Electron | 0 (nearly) | -1 |

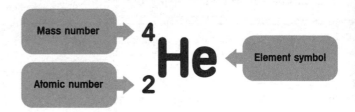

## Radioactive Decay

Radioactive isotopes (radioisotopes or radionuclides) are atoms with **unstable nuclei**. They may disintegrate and emit **radiation**. This is **radioactive decay**.

This decay can result in the formation of a **different atom** with a **different number of protons**.

Two examples of this are...
- **alpha** ($\alpha$) radiation
- **beta** ($\beta$) radiation.

There is also **gamma ($\gamma$) radiation**. However, unlike alpha and beta, gamma emissions have no effect on the structure of the nucleus.

## Alpha Decay

In alpha decay, the original atom decays by ejecting an alpha particle from the nucleus.

An alpha particle is a **helium nucleus** – a particle made up of **two protons** and **two neutrons**.

A new atom is formed with $\alpha$ decay.

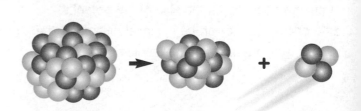

Unstable nucleus          New nucleus          $\alpha$ particle

# Radiation

## Beta Decay

In beta decay, the original atom decays by changing a **neutron** into a **proton** and an **electron**.

The newly formed high-energy electron ejected from the nucleus, is a β particle.

A new atom is formed with β decay.

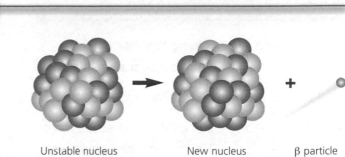

Unstable nucleus    New nucleus    β particle

## Ionisation

**Radioactive particles** can collide with **neutral** atoms or molecules. Electrons will be knocked out of their structure and they will become **charged**.

These charged particles are called **ions**. **Alpha** and **beta** radiation are known as **ionising radiation**.

This type of radiation can damage molecules in healthy cells. This can result in the death of the cell.

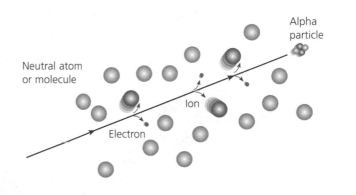

Alpha particle

Neutral atom or molecule

Ion

Electron

## Background Radiation

Background radiation is not harmful to our health as it occurs in very small doses.

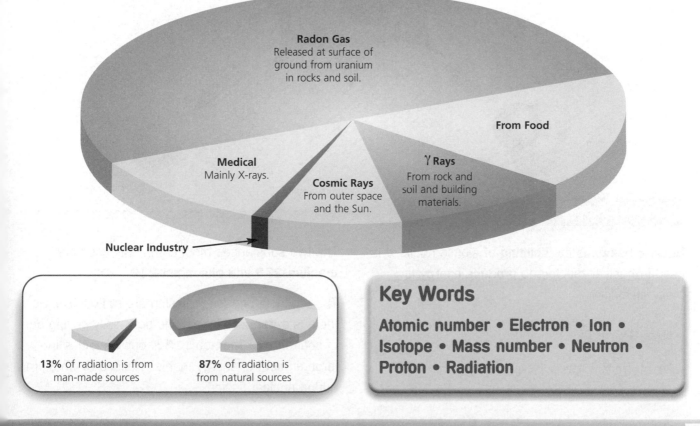

**Radon Gas** Released at surface of ground from uranium in rocks and soil.

**From Food**

**Medical** Mainly X-rays.

**Cosmic Rays** From outer space and the Sun.

**γ Rays** From rock and soil and building materials.

**Nuclear Industry**

**13%** of radiation is from man-made sources

**87%** of radiation is from natural sources

### Key Words

**Atomic number • Electron • Ion • Isotope • Mass number • Neutron • Proton • Radiation**

# Nuclear Fusion and Fission

## Nuclear Fusion

**Nuclear fusion** is the **joining together** of two or more atomic nuclei to form a **larger atomic nucleus**.

A lot of energy is needed for the nuclei to fuse.

A nuclear fusion reaction generally releases more energy than it uses. This makes it **self-sustaining**, i.e. some of the energy produced is used to drive further fusion reactions.

An example of nuclear fusion is the fusion of two heavy forms of hydrogen (deuterium and tritium).

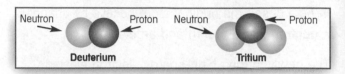

When they are forced together, the deuterium and tritium nuclei **fuse together** to form a **new helium atom** and an **unchanged neutron**.

Stars **release energy** by nuclear fusion. In the core of the Sun, **hydrogen** is **converted** to **helium** by fusion. This provides the energy to keep the Sun burning and allow life on Earth.

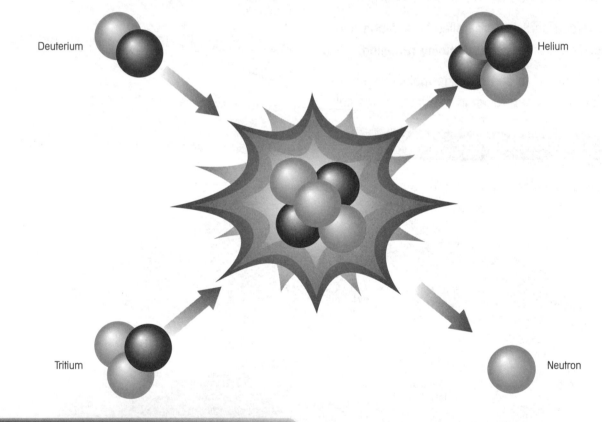

## Nuclear Fission

**Nuclear fission** is the **splitting** of atomic nuclei. It is used in nuclear reactors to **produce energy** to make electricity.

### Key Words

**Nuclear fission • Nuclear fusion**

The two substances most commonly used are **uranium-235** and **plutonium-239**.

The products of nuclear fission are **radioactive**. So, there is a **danger** of reactions potentially getting out of control (e.g. Chernobyl). But, one **benefit** is the large amount of energy that is released by an atom during nuclear fission.

## Nuclear Fission On a Small Scale

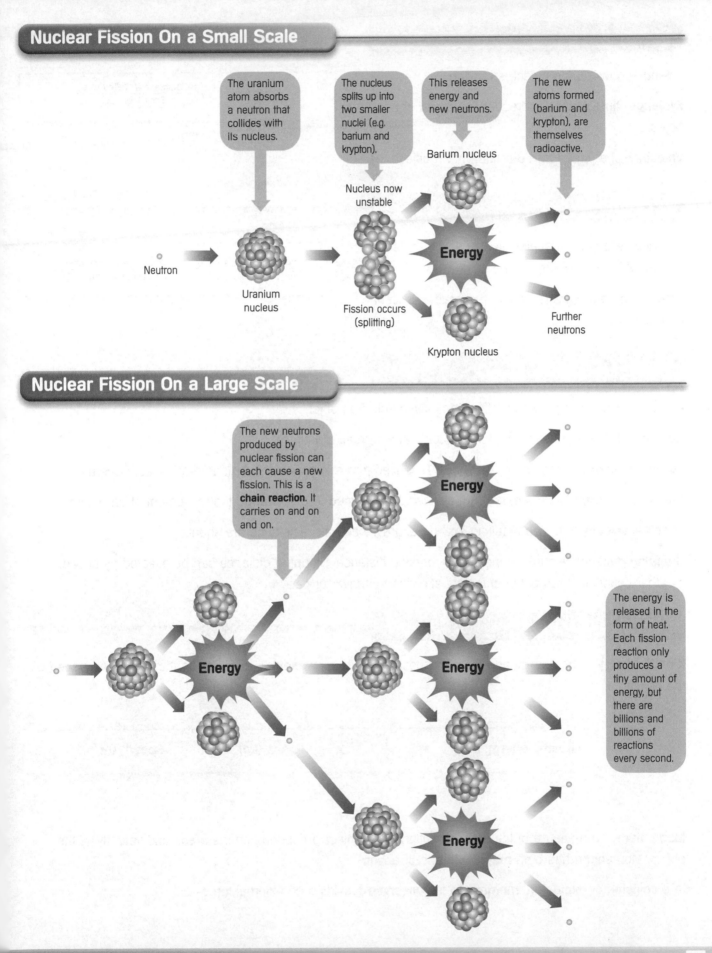

The uranium atom absorbs a neutron that collides with its nucleus.

The nucleus splits up into two smaller nuclei (e.g. barium and krypton).

This releases energy and new neutrons.

The new atoms formed (barium and krypton), are themselves radioactive.

Barium nucleus

Nucleus now unstable

**Energy**

Neutron

Uranium nucleus

Fission occurs (splitting)

Krypton nucleus

Further neutrons

## Nuclear Fission On a Large Scale

The new neutrons produced by nuclear fission can each cause a new fission. This is a **chain reaction**. It carries on and on and on.

**Energy**

**Energy**

**Energy**

**Energy**

The energy is released in the form of heat. Each fission reaction only produces a tiny amount of energy, but there are billions and billions of reactions every second.

# Unit 2 Summary

## Speed

**Speed** = how fast an object is moving.

**Distance–time graphs** represent the **speed** of an object.

**Velocity** = the **speed and direction** of an object.

$$\text{Speed (m/s)} = \frac{\text{Distance travelled (m)}}{\text{Time taken (s)}}$$

## Acceleration

The **acceleration** or **deceleration** of an object is the rate at which its **velocity changes**.

**Velocity–time graphs** represent the **acceleration** of an object.

$$\text{Acceleration (m/s}^2\text{)} \text{ (or deceleration)} = \frac{\text{Change in velocity (m/s)}}{\text{Time taken for change (s)}}$$

## Forces

**Forces** = pushes or pulls that affect the movement of an object.

**Resultant force** = the sum of all the forces that act on an object.

**Equal** and **opposite** forces are **balanced**. Balanced forces keep the movement of an object **constant**.

**Unequal** and **opposite** forces are unbalanced. **Unbalanced** forces **change** the movement of an object.

**Terminal velocity** is a steady falling speed, i.e. gravity and air resistance are equal.

**Stopping distance** = thinking distance + braking distance. Stopping distance can be affected by speed, weather conditions and external factors affecting the driver or car.

## Kinetic Energy

**Kinetic energy** = the energy an object has because of its **movement**. It depends on the **mass** and **speed** of the object.

**(HT)**

$$\text{Kinetic energy (J)} = \frac{1}{2} \times \text{Mass (kg)} \times \text{Speed}^2 \text{ (m/s)}^2$$

**Momentum** = a measure of the state of motion of an object. It depends on the **mass** and **velocity** of the object. Momentum has both **magnitude** and **direction**.

In a **collision** or **explosion**, **momentum** is **conserved** providing no external forces act.

## Static Electricity

Static electricity is created when two insulating materials are rubbed against each other.

Material minus electrons = positively charged

Material plus electrons = negatively charged

Like charges **repel**. Different charges **attract**.

Static electricity can be discharged by connecting it to earth with a conductor. Can be used in smoke precipitators and photocopiers.

## Circuits, Current and Power

The amount of electric current that flows through a component depends on…
* **potential difference** across component
* resistance of component.

Greater **potential difference** across a component = greater **current** through the component.

**Direct current** ➡ flows in the same direction. **Alternating current** ➡ constantly changes direction.

**Circuit breakers** and **fuses** are safety devices designed to **break** a circuit if the current becomes **too high**.

Electrical energy can be transformed into other types of energy. The **rate** of this transformation is the **power**.

HT

| Charge (coulomb, C) | = | Current (amp, A) | X | Time (second, s) |

## Atoms

Atom = protons + neutrons + electrons. Atom has no overall charge.

**Atomic number** = number of protons. **Mass number** = number of protons and neutrons.

## Radiation

**Background radiation** is not harmful.

**Ions** are **charged particles** caused by electrons being knocked out of atoms.

**Alpha** and **beta radiation** is **ionising** radiation.

## Nuclear Fission and Fusion

**Nuclear fusion** = joining of atomic nuclei.

**Nuclear fission** = splitting of atomic nuclei.

# Unit 2 Practice Questions

**1** This question is about speed. You may find this formula useful: $\text{Speed (m/s)} = \dfrac{\text{Distance travelled (m)}}{\text{Time taken (s)}}$

Match the letters A, B, C and D with the statements 1 to 4 below.

**A** 42                                          **B** 156.97

**C** 1.60                                        **D** 10.52

**1** What is the speed, in m/s, of a runner who travels a distance of 100 metres in 9.5 seconds? ☐

**2** A car travels a distance of 5023m at a speed of 32m/s. How many seconds does the car take to travel the distance? ☐

**3** It takes 3 seconds for an arrow to hit a target. If the arrow travels at a speed of 14m/s, how many metres from the target is the archer standing? ☐

**4** It takes a girl 19 minutes to walk to school and the school is 1827 metres away from her house. What is her walking speed, in m/s? ☐

**2** This question is about forces.

**a)** In what unit are forces measured? ........................................................

**b)** With what symbols are forces represented on diagrams? ........................................................

**c)** If a car travels at a constant speed, are the forces acting on the car balanced or unbalanced?

........................................................

**3** This question is about components in a circuit.

**a)** Explain what would happen to the resistance if the temperature of a thermistor increases.

........................................................

**b)** Sketch a diagram of a light dependent resistor.

**4** Complete the table about the mass and charges of the parts of an atom.

| Atomic Particle | Relative Mass | Relative Change |
|---|---|---|
| **a)** .................... | 1 | +1 |
| Neutron | **b)** .................... | 0 |
| **c)** .................... | 0 (nearly) | **d)** .................... |

**5** This question is about how a circuit breaker and a fuse work.

Match the letters A, B, C and D with the gaps 1 to 4 in the table.

**A** Circuit is broken

**B** Strength of electromagnet increases

**C** No current flows

**D** Current larger than current rating of fuse

| Circuit Breaker | Fuse |
|---|---|
| Current becomes too high | 2 ............................... |
| 1 ............................... | Fuse wire melts |
| Pair of contacts are pulled apart | Circuit is broken |
| 3 ............................... | 4 ............................... |
| Appliance or user is protected | Appliance or user is protected |

**6** The graph below shows the velocity of a motorbike.

**a)** In terms of speed, describe how the motorbike is travelling at the following stages:

**i)** Stage A

.....................................

**ii)** Stage C

.....................................

**iii)** Stage E

.....................................

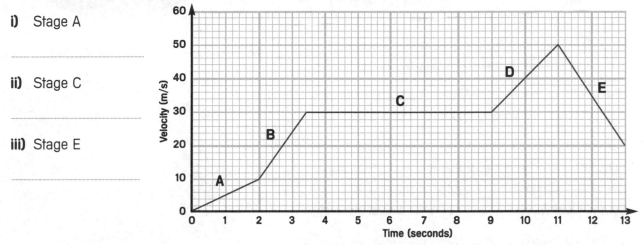

**b)** Calculate the acceleration at Stage D. Show your workings.

...............................................................................................................

...............................................................................................................

...............................................................................................................

**7** Calculate the momentum of a car that has a mass of 1700kg and is travelling at a velocity of 45m/s.

...............................................................................................................

...............................................................................................................

...............................................................................................................

# Moments

## Moments

Forces can be used to turn objects about a **pivot**. The **turning effect** of a force is called the **moment**.

For example, a **spanner** being used to unscrew a wheel nut would exert a **moment on the nut**.

You can **increase a moment** in two ways:
- **increase the force** applied
- **increase the perpendicular distance** between the line of action of the force and the pivot.

You can calculate the size of the moment by using this formula:

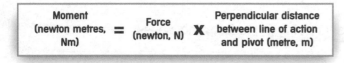

| Moment (newton metres, Nm) | = | Force (newton, N) | X | Perpendicular distance between line of action and pivot (metre, m) |

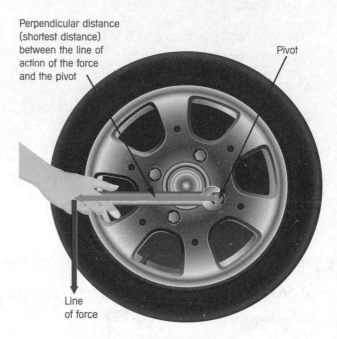

Perpendicular distance (shortest distance) between the line of action of the force and the pivot

Pivot

Line of force

## Centre of Mass

The **centre of mass** (C of M) of an object is the **point** through which the **whole weight** (W) of the **object acts**.

For example, if you balance an object on the end of your finger, the centre of mass is the point at which the object **balances**. The centre of mass of a **symmetrical object** is found along the **axis of symmetry**.

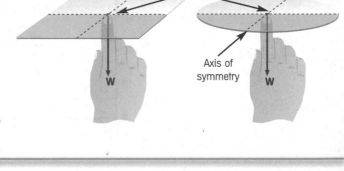

Centre of mass

Axis of symmetry

W

W

## Finding the Centre of Mass

A suspended object will always come to rest with its centre of mass directly **below** the point of suspension.

You can find the centre of mass of a sheet of material by this method:

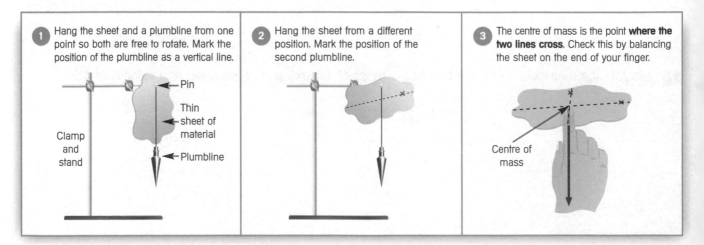

**1** Hang the sheet and a plumbline from one point so both are free to rotate. Mark the position of the plumbline as a vertical line.

Pin

Thin sheet of material

Clamp and stand

Plumbline

**2** Hang the sheet from a different position. Mark the position of the second plumbline.

**3** The centre of mass is the point **where the two lines cross**. Check this by balancing the sheet on the end of your finger.

Centre of mass

## Law of Moments

When an object is balanced, or isn't turning, there is a balance between…

- the **total moments** of the forces turning the object in a **clockwise direction**
- the **total moments** of the forces turning the object in an **anticlockwise direction**.

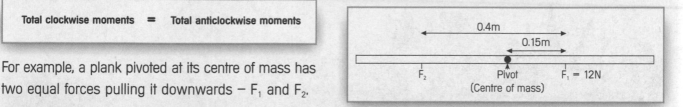

| Total clockwise moments | **=** | Total anticlockwise moments |
|---|---|---|

For example, a plank pivoted at its centre of mass has two equal forces pulling it downwards – $F_1$ and $F_2$.

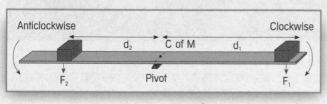

The object is **balanced** so total clockwise moments **are equal** to total anticlockwise moments.
So, $F_1 \times d_1 = F_2 \times d_2$.

### Example

A plank is pivoted at its centre of mass and has balanced forces acting. Calculate $F_2$.

Total clockwise moments = Total anticlockwise moments

$$12N \times 0.15m = F_2 \times (0.4 - 0.15)m$$

$$\text{So, } F_2 = \frac{12N \times 0.15m}{0.25m}$$

$$= \textbf{7.2N}$$

## Stability

An object will topple (fall over) if the **line of action of its weight** (the force) **lies outside its base**.

The **weight** of the object causes a **turning effect** which makes the object topple.

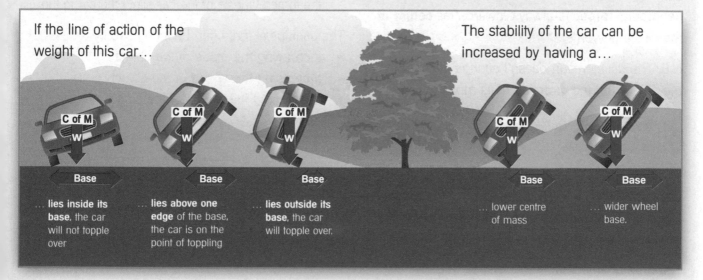

If the line of action of the weight of this car…

… **lies inside its base**, the car will not topple over

… **lies above one edge** of the base, the car is on the point of toppling

… **lies outside its base**, the car will topple over.

The stability of the car can be increased by having a…

… lower centre of mass

… wider wheel base.

## Key Words

Centre of mass • Force • Moment • Pivot

# Centripetal Force and Gravity

## Motion in a Circle

Many objects move in **circular**, or near circular, paths. For example…

- a rubber ball spun on a piece of string
- spinning rides at fairgrounds
- a car turning
- the Earth and other planets in orbit around the Sun.

## Centripetal Force

An object moving in a circular path is **continuously accelerating** towards the centre of the circle.

The **resultant force** causing this acceleration is called the **centripetal force**. The direction of **centripetal forces** is always towards the **centre of the circle** (inwards).

*N.B. The acceleration doesn't change the speed of the object, but the **direction** of its motion.*

Several different forces can act in this way:

- A whirling ball – the centripetal force that keeps the ball moving is provided by the **tension force** in the string.
- A turning car – the **frictional force** acting on the turned wheels of the car allow the car to turn.

The centripetal force can be increased by…

- **increasing the mass** of the object
- **increasing the speed** of the object
- **decreasing the radius** of the circle.

## Gravity and our Solar System

The Earth, Sun, Moon and all other objects attract each other with a force called gravity. **The bigger the mass** of the object, the **bigger the force of gravity** between them.

The orbit of any planet is an **ellipse** (slightly squashed circle) with the Sun close to the centre.

Gravitational force provides the centripetal force for the elliptical orbits of planets and satellites. For example…

- the Moon and artificial satellites around the Earth
- Earth and other planets around the Sun.

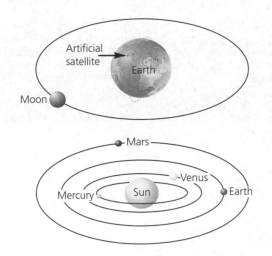

# Centripetal Force and Gravity

## Gravity and Orbiting Speed

As the **distance** between two objects **increases**, the **force of gravity** between them proportionally **decreases**.

For example, if the distance is...

- **doubled**, the force becomes $\frac{1}{4}$ of the original force
- **trebled**, the force becomes $\frac{1}{9}$ of the original force.

To stay in orbit at a particular distance, objects must orbit at a **particular speed**. This balances the gravitational force.

So, the further away an orbiting object is, the longer it takes to make a complete orbit.

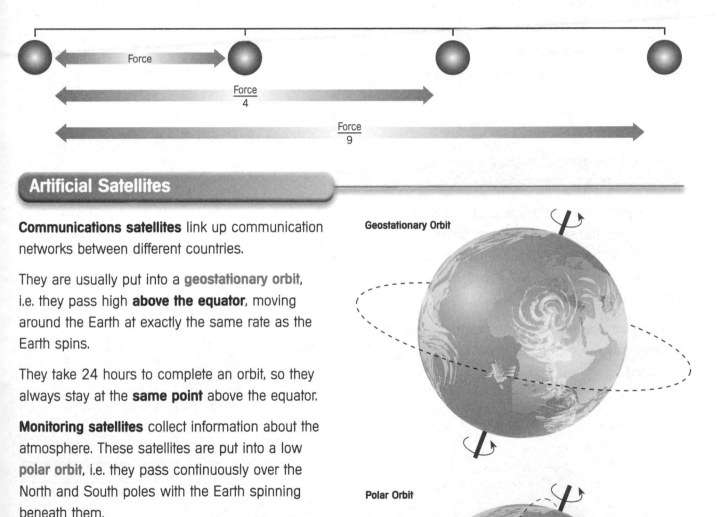

## Artificial Satellites

**Communications satellites** link up communication networks between different countries.

They are usually put into a **geostationary orbit**, i.e. they pass high **above the equator**, moving around the Earth at exactly the same rate as the Earth spins.

They take 24 hours to complete an orbit, so they always stay at the **same point** above the equator.

**Monitoring satellites** collect information about the atmosphere. These satellites are put into a low **polar orbit**, i.e. they pass continuously over the North and South poles with the Earth spinning beneath them.

They orbit and scan the Earth several times every day from a much closer range than a geostationary satellite.

**Geostationary Orbit**

**Polar Orbit**

# Reflection and Refraction

## Reflection of Light

When light strikes a surface it changes direction. This is called **reflection**.

The **normal line** is perpendicular to the reflecting surface at the **point of incidence**. The normal line is used to calculate the angles of incidence and reflection.

| Angle of incidence | = | Angle of reflection |
|---|---|---|

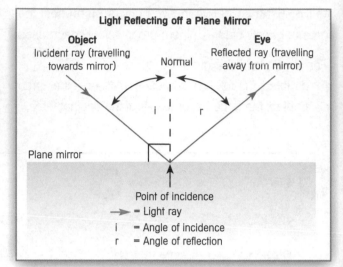

**Light Reflecting off a Plane Mirror**

**Object** — Incident ray (travelling towards mirror)

Normal

**Eye** — Reflected ray (travelling away from mirror)

i    r

Plane mirror

Point of incidence

→ = Light ray
i = Angle of incidence
r = Angle of reflection

## Refraction of Light

When light crosses an interface (a boundary between two transparent media of different densities) it changes direction unless it meets the boundary at an angle of 90° (along the normal). This is called **refraction**.

A **triangular prism** doesn't have parallel sides, so a light ray travelling through it is **deviated** (changes direction).

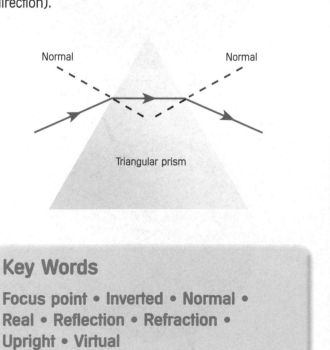

Normal      Normal

Triangular prism

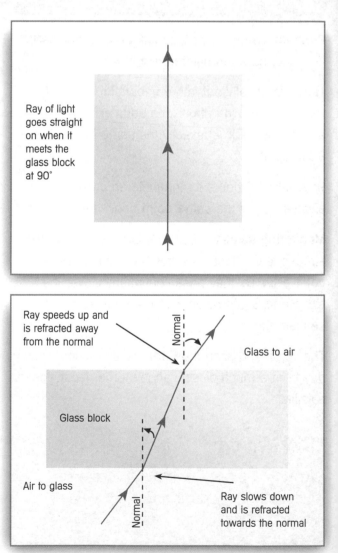

Ray of light goes straight on when it meets the glass block at 90°

Ray speeds up and is refracted away from the normal

Normal

Glass to air

Glass block

Air to glass

Normal

Ray slows down and is refracted towards the normal

Normal

## Key Words

**Focus point • Inverted • Normal • Real • Reflection • Refraction • Upright • Virtual**

## Images Produced by Mirrors

An image is a representation of an object and it is defined by its size or position relative to the object. For example, whether it's...

- **upright** or **inverted**
- **real** or **virtual**.

Different mirrors produce different types of images.

1 **A plane mirror** forms an image that's the **same size** as the object. It is **upright** and **laterally inverted** (faces the opposite way to the object).

The rays of light reaching the eye **appear** to come directly from the image. But, because the rays of light **don't actually** come from the image, it's described as a **virtual image**.

2 A **convex** (curving outwards) mirror forms an image that **appears** to be formed at a single point behind the mirror. The image is **smaller than the object** and is **virtual and upright**.

3 A concave mirror forms an image inside its curve at the point where all the rays of light **converge** (come together). This is called the **focus point** (F).

Concave mirrors produce different images **depending on the distance** of the object from the mirror.

If the distance between the object and the mirror is **less** than the distance between the mirror and F, the image produced is **virtual, upright** and **larger** than the object.

If the distance between the object and the mirror is **greater** than the distance between the mirror and F, the image is...

- **inverted** (upside down) and **laterally inverted**
- **real** (the rays of light actually meet rather than just appearing to)
- of **varying size**, depending on the distance from the mirror (the greater the distance, the smaller the object).

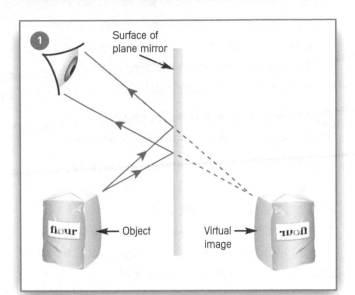

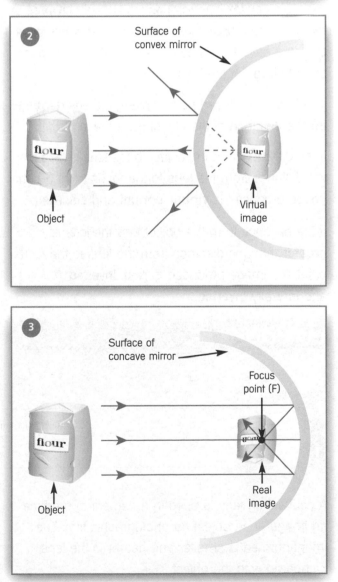

# Lenses

## Images Produced by Lenses

A lens is a piece of transparent material which **refracts** light rays. There are two types of lens:

- **diverging** (concave) – thinnest at its centre
- **converging** (convex) – thickest at its centre.

The different lenses have a different curvature, so parallel rays of light **pass through them differently**.

## Diverging and Converging Lenses

A double **concave** lens refracts rays of light outwards at the two curved boundaries so they **appear to come** from one point, the focus (F). Only the ray which meets the lens at 90° will pass straight through the lens. The image produced by a diverging lens is **virtual** and **upright**.

A double **convex** lens **refracts** rays of light **inwards** at the two curved boundaries so they **meet at the focus**. Only the ray which meets the lens at 90° will pass straight through the lens.

The image produced by a converging lens **depends on the distance** of the object from the lens.

If the distance from the object to the lens is **less** than the distance from the lens to the focus point (F), the image produced is **virtual**, **upright** and **enlarged**.

If the distance from the object and the lens is **greater** than the distance from the lens to the focus point, the image produced is **real**, **inverted** and **laterally inverted**.

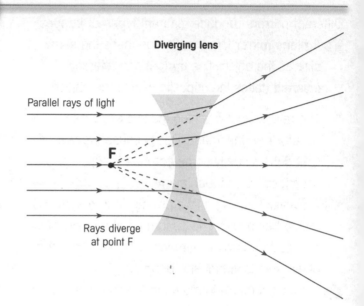

**Diverging lens**

Parallel rays of light

F

Rays diverge at point F

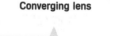

**Converging lens**

Parallel rays of light

Rays converge at point F

F

## The Camera

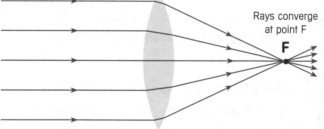

Distant object

Rays to form image

Photographic film

Converging lens

Image

A converging lens is used in a camera to produce an image of an object on photographic film. The image formed is smaller and nearer to the lens, compared with the object.

# Sound

## Sound Waves

Sounds travel as waves. The waves are produced when something mechanical vibrates backwards and forwards. The quality of a note depends on the waveform.

Sound can't travel through a vacuum. It can be…
- **reflected** off hard surfaces to produce echoes
- **refracted** when it passes into a different medium or substance.

## Frequency and Pitch

The **frequency** of a sound wave is the **number of vibrations** produced in one second.

Frequency is measured in **hertz** (Hz). Humans can hear sounds in the range of 20–20 000Hz.

The frequency affects the **pitch** of the sound.

As the frequency is increased, the sound becomes higher pitched.

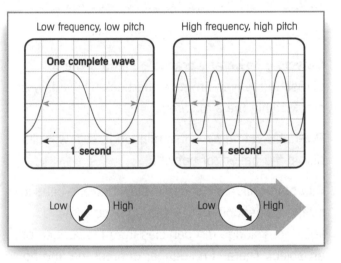

## Amplitude and Loudness

**Amplitude** is the **peak of movement** of the sound wave from its rest point.

The amplitude affects the **loudness** of the sound.

As the amplitude is increased, the sound becomes louder.

### Key Words

**Amplitude • Converging • Diverging • Frequency • Pitch • Reflection • Refraction**

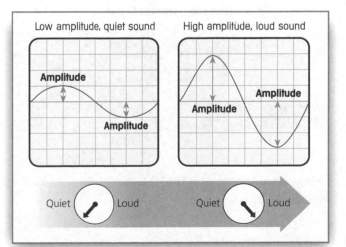

# Ultrasound

## Ultrasound

**Ultrasound** is sound waves of **frequencies greater** than 20 000Hz, i.e. above the upper limit of the hearing range for humans.

Electronic systems produce electrical oscillations which are used to generate the ultrasonic waves.

As ultrasonic waves pass from one medium or substance into another, they are **partially reflected** at the boundary. The time taken for these reflections is a measure of **how far away** the boundary is.

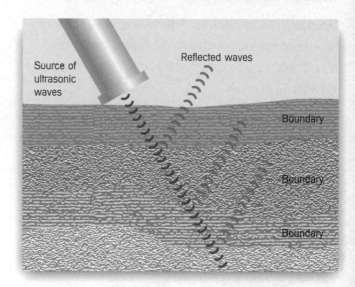

## Uses of Ultrasound

Ultrasound can be used in many ways.

**Detecting flaws and cracks** – some of the ultrasound waves are reflected back by the flaw or crack within the structure.

**Pre-natal scanning** – this method is safe with little risk to the patient or baby.

**Cleaning delicate objects** – the vibrations caused by the ultrasound waves can be used within a liquid to dislodge dirt particles from the surface of an object. There is no danger of breakage and no need to take the object apart.

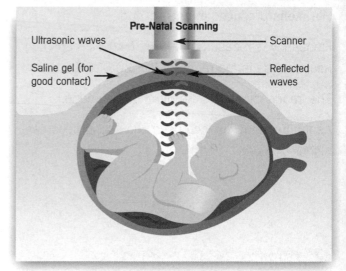

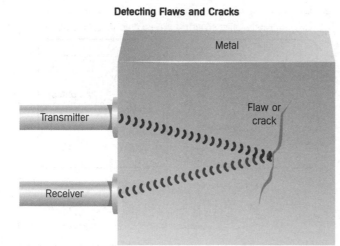

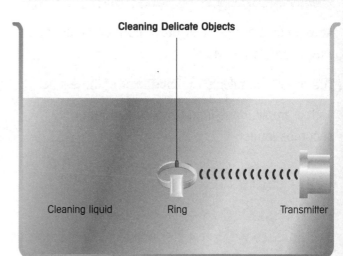

## The Motor Effect

The motor effect uses **current to produce movement**.

When a **conductor** (wire) carrying an electric current is placed in a **magnetic field**, the magnetic field formed around the wire interacts with the permanent magnetic field. This causes the wire to experience a force which makes it move.

*N.B. The wire will not experience a force if it's parallel to the magnetic field.*

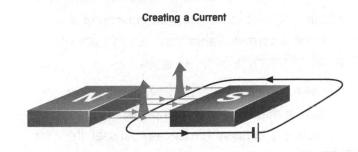

Creating a Current

| The size of the force on the wire can be **increased** by... | The direction of the force on the wire can be **reversed** by... |
| --- | --- |
| • **increasing the size of the current** (e.g. having more cells) | • reversing the **direction of flow** of the current (e.g. turning the cell around) |
| • **increasing the strength** of the **magnetic field** (e.g. having stronger magnets). | • reversing the **direction of the magnetic field** (e.g. swapping the magnets around). |

## Direct Current Motor

Electric motors use the principle of the motor effect and they form the basis of a huge range of electrical devices both inside and outside of the home.

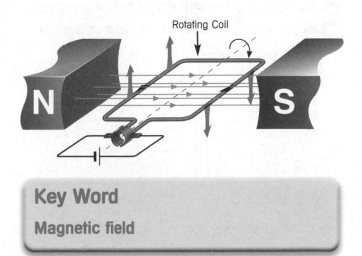

Rotating Coil

1. As a current flows through the coil, a magnetic field is formed around the coil. This creates an electromagnet.
2. The magnetic field interacts with the permanent magnetic field which exists between the two poles, North and South.
3. A force acts on both sides of the coil, so the coil rotates and a very simple motor is generated.

### Key Word

**Magnetic field**

# Generating Electrical Current

## Electromagnetic Induction

Electromagnetic induction uses **movement** to produce a **current**. **Generators** use this effect to produce electricity.

If a wire, or coil of wire, cuts through the lines of force of a magnetic field (or vice versa), a potential difference is **induced** (produced) between the ends of the wire. If the wire is part of a complete circuit, a current will be induced.

Moving the magnet into the coil induces a current in one direction. A current can be induced in the **opposite direction** in two ways:
1. Moving the magnet out of the coil.
2. Moving the other pole of the magnet into the coil.

N.B. The same effects work if the magnetic field is stationary and the coil is moved. But remember, if there's no movement of magnet or coil, no current is induced.

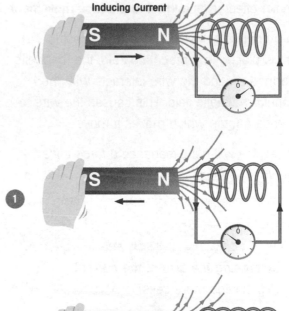

## Increasing Potential Difference

The size of the induced **potential difference** can be increased by **increasing** the **area of the coil**.

Potential difference can also be increased by the following methods:
1. By increasing the **speed of movement** of the magnet or the coil.
2. By increasing the **strength** of the magnetic field.
3. By increasing the **number of turns** on the coil.

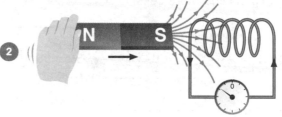

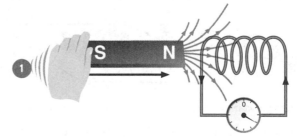

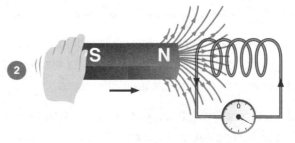

## Key Words

Potential difference • Transformer

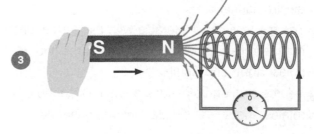

## Transformers

A **transformer** changes electrical energy from one **potential difference** to another potential difference. They are made of **two coils**, called the **primary** and **secondary** coils, wrapped around a **soft iron core**.

An **alternating potential difference** across the primary coil causes an **alternating current to flow** (input). This alternating current creates a **continually changing magnetic field** in the iron core, which creates an **alternating potential difference** across the ends of the secondary coil (output).

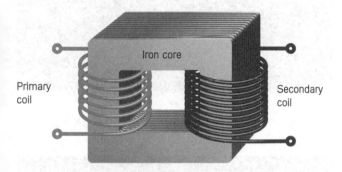

Iron core

Primary coil

Secondary coil

(HT) You can calculate the size of the potential difference across the primary and secondary coils by using this formula:

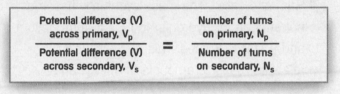

| Potential difference (V) across primary, $V_p$ | = | Number of turns on primary, $N_p$ |
|---|---|---|
| Potential difference (V) across secondary, $V_s$ | | Number of turns on secondary, $N_s$ |

**Example**

A transformer has 200 turns on the primary coil and 800 turns on the secondary coil. If a potential difference of 230V is applied to the primary coil, what is the potential difference across the secondary coil?

$$\frac{V_p}{V_s} = \frac{N_p}{N_s}$$

$$\frac{230V}{V_s} = \frac{200}{800}$$

$$\frac{230 \times 800}{200} = \textbf{920 volts}$$

## Step-Up and Step-Down Transformers

**A step-up transformer** has more turns in the secondary coil than the primary coil. The potential difference leaving the secondary coil is **greater** than that across the primary coil.

**A step-down transformer** has fewer turns in the secondary coil than the primary coil. The potential difference leaving the secondary coil is **less** than that across the primary coil.

Step-up and step-down transformers are used in the National Grid to ensure the efficient transmission of electricity.

**National Grid Process**

Power Station 25 000V

Step-Up Transformer

Power Lines 400 000V

Step-Down Transformer

Houses, shops, etc. 230V

**A Step-Up Transformer**

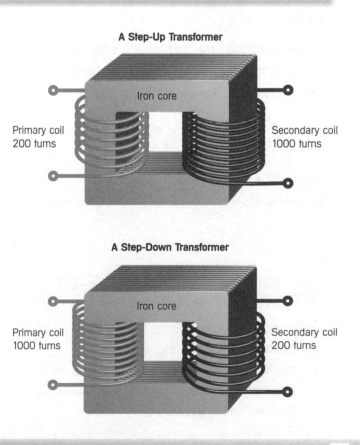

Iron core

Primary coil 200 turns

Secondary coil 1000 turns

**A Step-Down Transformer**

Iron core

Primary coil 1000 turns

Secondary coil 200 turns

# Stars

## Formation of Stars

This is how a star is formed:

1. **Gravitational attraction** pulls clouds of dust, rock and gas (nebulae) together.
2. Heat is created. The mass eventually becomes hot enough for hydrogen to fuse to form helium, and a star is formed. This nuclear fusion releases massive amounts of energy and produces all naturally occurring elements.
3. Smaller masses may also be formed and be attracted by larger masses to become planets.

Stars use hydrogen as their energy source, so they can release energy for millions of years. Our Sun is believed to be 5 billion years old and only half-way through its life. It is made up of approximately 74% hydrogen and 24% helium, with traces of heavier elements.

## Our Galaxy and the Universe

Our Sun is one of the many billions of stars in our galaxy, the **Milky Way**. The stars in a galaxy are often millions of times further apart than the planets in a Solar System.

The Milky Way is one of at least a billion galaxies in the Universe. Galaxies are often millions of times further apart than the stars within a galaxy.

### Key Words

Gravity • Radiation

Not to scale

Our Sun

Our Sun

Our galaxy the Milky Way

Our galaxy

The Universe

## The Life Cycle of a Star

A star remains stable during its life period due to the **balance** of two forces:
- the force of gravity pulling the star inwards
- **huge temperatures** (radiation pressure) within the star acting **outwards**.

Towards the end of the star's life, different processes may occur **depending on the mass** of the star.

## Smaller Stars

This is the life cycle of stars the size of our Sun:

1 Star expands to become a **red giant**.

2 Red giant then cools down and eventually collapses to become a **white dwarf** (with a density millions of times greater than any matter on Earth).

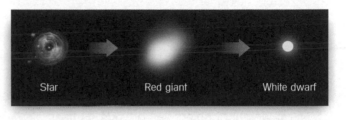

Star · Red giant · White dwarf

## Bigger Stars

This is the life cycle of stars at least four times bigger than our Sun:

1 Star expands enormously to become a **red supergiant**.

2 Star then shrinks rapidly and explodes – this is called a **supernova**. The explosion releases massive amounts of energy, dust and gas into space.

A **medium-sized star** (ten times bigger than our Sun) will then form a **neutron star**. This is the core of the star that remains after the explosion. A neutron star is made only of neutrons and is very dense.

**Large stars** (greater than ten times the size of our Sun) will leave behind **black holes**, where the matter is so dense and the gravitational field so strong that nothing, not even electromagnetic radiation, can escape from it.

Black holes can only be observed indirectly through their effects on their surroundings, e.g. the X-rays emitted when gases from a nearby star spiral into a black hole.

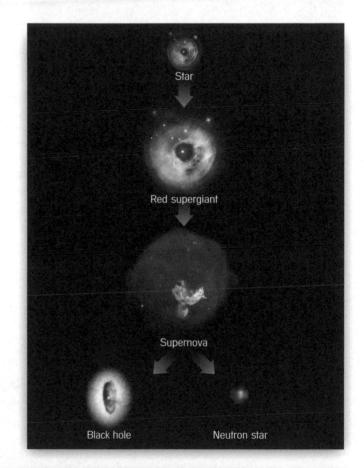

Star
Red supergiant
Supernova
Black hole · Neutron star

## HT Recycling Stellar Material

Stars need hydrogen as a fuel to undergo nuclear fusion. Helium is produced as a result.

However, during fusion, hydrogen and helium can also fuse together to produce nuclei of heavier elements including iron.

As a star comes to the end of its life and explodes, its elements are distributed throughout the Universe.

So, a large variety of different elements are circulated in the Universe, not just hydrogen.

These elements can be recycled in the formation of new stars or planets. Atoms of heavier elements are present in the inner planets of the Solar System which leads us to believe that the Solar System was formed from the material produced when earlier stars exploded.

# Unit 3 Summary

## Moments and Centre of Mass

**Moment** = turning effect of a force.

(HT) A **balanced** or **stationary** object ➡ total **clockwise** moments **=** total **anticlockwise** moments.

If weight of object lies **inside its base** ➡ object will not topple.

If weight of object lies **outside its base** ➡ object will topple.

## Centre of Mass

**Centre of mass** = the point through which the whole weight of an object acts.

Centre of mass of a symmetrical object ➡ axis of symmetry.

## Centripetal Force and Gravity

**Centripetal force** = resultant force which causes an object moving in a circular path to continuously accelerate towards the centre of the circle.

Centripetal force be increased by…
- increasing mass or speed of object
- decreasing radius of circle.

**Gravity** = force which allows objects to attract other objects.

**Gravitational force** = centripetal force that maintains the elliptical orbits of planets and satellites.

As the **distance** between two objects **increases**, the **force of gravity** between them proportionally **decreases**.

Communication satellites ➡ Geostationary orbit (above the equator).

Monitoring satellites ➡ Low polar orbit.

## Mirrors and Lenses

**Reflection** = change in direction of a light ray as it hits a surface.

**Refraction** = change in direction of a light ray as it crosses an interface.

**Plane mirror** ➡ image upright, laterally inverted, virtual, same size as object.

**Convex mirror / diverging lens** ➡ image virtual, upright, smaller than object.

**Concave mirror / converging lens** ➡ distance between object and mirror less than distance between mirror and F ➡ image virtual, upright, larger than object.

**Concave mirror / converging lens** ➡ distance between object and mirror less than distance between mirror and F ➡ image inverted, laterally inverted, real.

## Sound and Ultrasound

**Sound** = waves in the range 20–20 000Hz.

Increased frequency ➡ higher pitch. Increased amplitude ➡ louder sound.

**Ultrasound** = waves above the range 20 000Hz. Can be used in pre-natal scanning, for detecting flaws and cracks, and for cleaning.

## Motors and Electromagnetic Induction

**Motor effect** = current produces movement.

Size of force increased by increasing the…

- size of current
- strength of magnetic field.

Direction of force on wire reversed by reversing the direction of…

- flow of current
- magnetic field.

**Electromagnetic induction** = movement produces current.

Size of potential difference increased by increasing the…

- speed of movement of magnet or coil
- strength of magnetic field
- number of turns on coil or area of coil.

## Transformers

**Transformer** ➡ changes potential difference.

**Step-up transformer** = more turns in secondary coil than primary coil.

**Step-down transformer** = more turns in primary coil than secondary coil.

(HT)

$$\frac{\text{Potential difference across primary (V), } V_p}{\text{Potential difference across secondary (V), } V_s} = \frac{\text{Number of turns on primary, } N_p}{\text{Number of turns on secondary, } N_s}$$

## Stars

Stars are formed when gravity pulls clouds of dust, rock and gas together.

Our Sun = one of billions of stars in the Milky Way. **Milky Way** = one of billions of galaxies in the Universe.

**Stable star** = when gravity (pulling inwards) and radiation (pulling outwards) are equal.

Small stars ➡ red giant ➡ white dwarf

Large stars ➡ red supergiant ➡ supernova ➡ neutron star (x4 size of Sun)

supernova ➡ black hole (x10 size of Sun)

# Unit 3 Practice Questions

**1** What is the centre of mass of an object?

.................................................................................................................................................................

**2 a)** What is a **moment**? ...................................................................................................................................

**b)** The force applied to a spanner is 5N and the distance between the line of action of the force and the pivot is 0.25m. What is the moment?

.................................................................................................................................................................

.................................................................................................................................................................

**c)** If the moment of a spanner is 12Nm and the force applied is 5N, calculate the distance between the line of action of the force and the pivot.

.................................................................................................................................................................

.................................................................................................................................................................

**d)** If an object is balanced, or isn't turning, which two moments are balanced?

.................................................................................................................................................................

**3 a)** A force accelerates continuously towards the centre of a circle. What is the name of this force?

.................................................................................................................................................................

**b)** Name two forces that can act in this way.

**i)** ...................................................................... **ii)** ......................................................................

**4** Correctly label the diagram.

**A)** ...................................................................... **B)** ......................................................................

**C)** ...................................................................... **D)** ......................................................................

5 Complete the following diagrams to show what happens to rays of light when they hit a surface.

a)

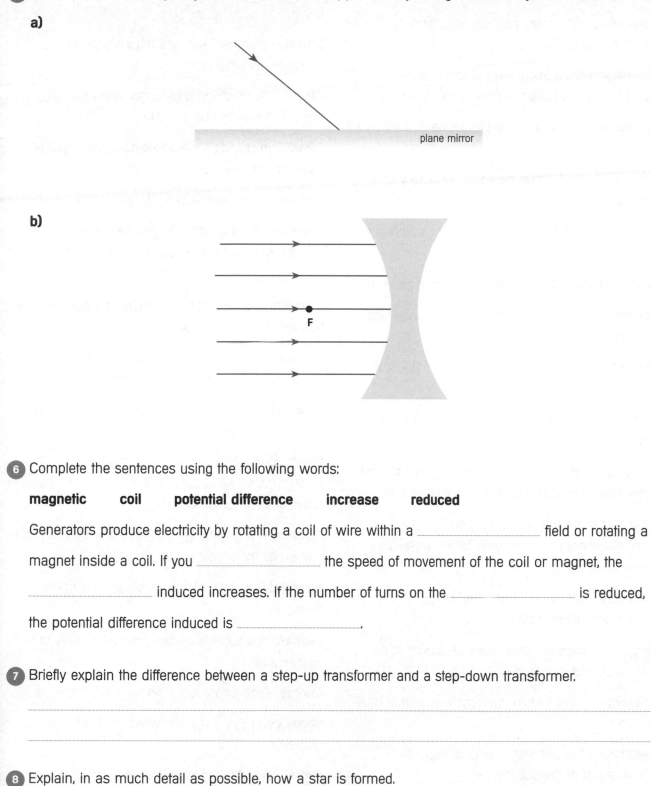

plane mirror

b)

F

6 Complete the sentences using the following words:

**magnetic     coil     potential difference     increase     reduced**

Generators produce electricity by rotating a coil of wire within a _____ field or rotating a

magnet inside a coil. If you _____ the speed of movement of the coil or magnet, the

_____ induced increases. If the number of turns on the _____ is reduced,

the potential difference induced is _____ .

7 Briefly explain the difference between a step-up transformer and a step-down transformer.

_____

_____

8 Explain, in as much detail as possible, how a star is formed.

_____

_____

_____

# Glossary of Key Words

**Absorption** – a substance's ability to absorb energy.

**Acceleration** – the rate at which an object increases in speed.

**Alternating current (a.c.)** – an electric current which changes direction of flow continuously.

**Amplitude** – the maximum disturbance caused by a wave.

**Analogue** – a signal that varies continuously in amplitude / frequency.

**Atom** – the smallest part of an element which can enter into a chemical reaction.

**Atomic number** – the number of protons in an atom.

**Attraction** – the drawing together of two materials with different charges.

**Centre of mass** – the point at which an object's mass is concentrated.

**Charged** – having an overall positive or negative electrical charge.

**Centripetal force** – the external force required to make an object follow a circular path at a constant speed.

**Circuit breaker** – a safety device which automatically breaks an electric circuit when it becomes overloaded.

**Conduction** – the transfer of heat energy without the substance itself moving.

**Energy** – the capacity of a physical system to do work; measured in joules (J).

**Conductor** – a substance that readily transfers heat or energy.

**Convection** – the transfer of heat energy through the movement of the substance.

**Converging lens** – a lens in which light rays passing through it are brought to a central point.

**Current** – the flow of electric charge through a conductor.

**Deceleration** – the rate at which an object slows down.

**Digital** – a signal that uses binary code to represent information.

**Diode** – an electrical device that allows electric current to flow in one direction.

**Direct current (d.c.)** – an electric current which flows in one direction.

**Distance** – the space between two points.

**Distance–Time graph** – a graph showing distance travelled against time taken; the line represents speed.

**Diverging lens** – a lens in which light rays passing through it are spread out.

**Earthed** – connecting the metal case of an electrical appliance to the earth pin of a plug.

**Efficiency** – the ratio of energy output to energy input, expressed as a percentage.

**Electromagnetic spectrum** – a continuous arrangement that displays electromagnetic waves in order of increasing frequency.

**Electron** – a negatively charged subatomic particle that orbits the nucleus.

**Electrostatic** – producing, or caused by, static electricity.

**Element** – a substance that consists of only one type of atom.

**Energy** – the ability to do work; measured in joules.

**Focus point** – the point at which all light rays converge.

**Force** – a push or pull acting on an object.

**Fossil fuel** – fuel formed in the ground, over millions of years, from the remains of dead plants and animals.

**Frequency** – the number of times something happens in 1 second; measured in hertz.

**Friction** – the resistive force between two surfaces as they move over each other.

**Fuse** – a thin piece of metal which overheats and melts to break an electric circuit if it's overloaded.

**Geostationary orbit** – an orbit which passes high over the equator.

**Gravity** – the force which attracts masses together.

**Half-life** – the time taken for half the atoms in a radioactive material to decay.

**Inverted** – upside down.

**Ion** – a particle that has a positive or negative electrical charge.

**Ionising power** – the ability to ionise other atoms.

**Isotopes** – atoms of the same element which contain different numbers of neutrons.

**Kinetic energy** – the energy possessed by an object due to its movement.

**Magnetic field** – a force that is present around a magnetic field.

**Mass** – the quantity of matter in an object.

**Mass number** – the total number of protons and neutrons present in an atom.

**Moment** – a turning force.

**Momentum** – a measure of the state of motion of an object as a product of its mass and velocity.

**National Grid** – a network of cables that carries electricity to industries and homes.

**Neutron** – a subatomic particle found in the nucleus which has no charge.

**Newton** – a measure of force.

**Non-renewable** – energy sources that can't be replaced in a lifetime.

**Normal line** – the line which is at 90° to the reflecting / refracting surface at the point of incidence.

**Nuclear** – non-renewable fuel.

**Nuclear fission** – the splitting of atomic nuclei.

**Nuclear fusion** – the joining together of atomic nuclei.

**Optical fibre** – a thin strand of glass / plastic that uses totally internally reflected light to carry information.

**Parallel circuit** – a circuit where there are two (or more) paths for the current to take.

**Pitch** – the frequency of vibration; how high or low a sound is; changes with frequency.

**Pivot** – an axis consisting of a short shaft that supports something that turns.

**Polar orbit** – an orbit which passes over the North and South poles.

**Power** – the rate of doing work; measured in watts.

**Potential difference / voltage** – the difference in electrical charge between two charged points.

**Proton** – a positively charged subatomic particle found in the nucleus.

**Radiation** – electromagnetic particles / rays emitted by a radioactive substance.

**Real image** – an image produced by rays of light meeting at a point.

**Red shift** – the shift of light towards the red part of the visible spectrum; shows that the Universe is expanding.

**Reflection** – a change in direction of a wave as it hits a surface.

**Refraction** – the change in direction and speed of a wave as it passes from one medium to another.

**Renewable** – energy sources that can be replaced.

# Glossary of Key Words

**Repulsion** – the pushing away of two materials with the same charge.

**Resistance** – opposition to the flow of an electric current.

**Resistor** – an electrical device that resists the flow of an electric current.

**Resultant force** – the combined effect of all the forces acting on an object.

**Series circuit** – a circuit where there is one path for the current to take.

**Speed** – the rate at which an object moves.

**Static electricity** – electricity produced by friction.

**Telescope** – a device that magnifies distant objects.

**Terminal velocity** – the constant maximum velocity reached by a falling object (i.e. gravitational force is equal to the frictional forces).

**Thermal energy** – heat energy.

**Thermistor** – a resistor whose resistance varies with temperature.

**Transfer** – to move energy from one place to another.

**Transform** – to change energy from one form into another, e.g. electrical energy to heat energy.

**Transformer** – an electrical device used to change the voltage of alternating currents.

**Transmission** – the sending of information or electricity over a communications line or circuit.

**Upright** – the correct way up.

**Velocity** – the speed at which an object moves in a particular direction.

**Velocity–Time graph** – a graph showing velocity against time taken; the line represents acceleration.

**Virtual image** – an image produced by rays of light appearing to meet at a point.

**Voltage / potential difference** – the difference in electrical charge between two charged points.

**Weight** – the gravitational force exerted on an object.

**Work** – the energy transfer that occurs when a force causes a body to move a certain distance.

## Physics Unit 1a

1. A 1; B 4; C 2; D 3.

2. A 4; B 1; C 3; D 2.

3. By a network of cables called the National Grid.

4. 75%

5. a) ii) Sound and light

   b) iii) Heat

   c) iv) 31% of the energy is wasted as sound and heat.

   d) ii) The energy is used up.

6. a) Light

   b) Time = 5 x 60 = 300 seconds

   max energy = 45 x 300

   = 13 500 joules

c) **Any two from**: Ideal for remote places; Don't need to be connected to mains; Produce free, clean electricity once set up; Running costs are low; Renewable; Good for small amounts of electricity.

7. a) **Any two from**: Coal; Oil; Gas.

   b) **Any three from**: Wind turbines; Solar cells / panels; Hydro-electric; Tidal barrage; Nodding duck; Geothermal.

8. A 4; B 3; C 2; D 1.

## Physics Unit 1b

1. A 2; B 4; C 3; D 1.

2. A 2; B 4; C 1; D 3.

3. a) **Any two from**: No change in signal quality; More information can be transmitted; Can be processed by a computer; Not affected by background noise.

   b)

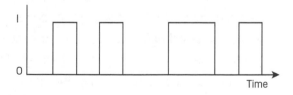

   c) Because digital signals are either on (1) or off (0) and don't pick up interference. When amplified, the quality of digital signals is retained. There is deterioration in analogue sounds.

4. A 3; B 1; C 4; D 2.

5. A 2; B 1; C 3; D 4.

6. a) The greater the thickness of the paper the more radiation is absorbed. The detector picks up less radiation. A signal is sent to the rollers to make them move closer together.

   b) **Any two from**: Sterilisation; Treating cancer; Tracers.

7. **Any one from**: The time taken for half the number of nuclei of the isotope to halve; The time taken for the count rate of the isotope to halve.

8. A 1; B 2; C 3; D 4.

# Answers to Practice Questions

## Physics Unit 2

**1.** **1** D, **2** B, **3** A, **4** C.

**2.** **a)** Newtons

   **b)** Arrows

   **c)** Balanced

**3.** **a)** The resistance would decrease.

   **b)**

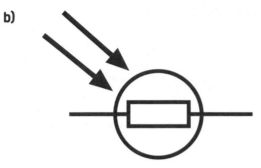

**4.** **a)** Proton

   **b)** 1

   **c)** Electron

   **d)** -1

**5.** **1** B, **2** D, **3** A, **4** C.

**6.** **a)** **i)** Speeding up / accelerating.

     **ii)** Steady or constant speed.

     **iii)** Slowing down / decelerating.

   **b)** Acceleration $= \dfrac{\text{Change in velocity}}{\text{Time taken}} = \dfrac{20m/s}{2s} = 10m/s^2$.

**7.** Momentum = Mass x Velocity
$$= 1700 \times 45 = 76\ 500kg\ m/s$$

## Physics Unit 3

**1.** The point through which the whole weight of the object acts.

**2.** **a)** A turning force.

   **b)** Moment = Force x Distance
           = 5 x 0.25
           = 1.25Nm

   **c)** Distance $= \dfrac{\text{Moment}}{\text{Force}} = \dfrac{12}{5} = 2.4m$

   **d)** Total clockwise moments and total anticlockwise moments.

**3.** **a)** Centripetal force.

   **b)** **Accept two from**: tension, gravity, friction.

**4.** **A** – incident ray, **B** – normal,
   **C** – glass block, **D** – refracted ray.

**5. a)**

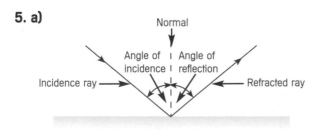

Normal

Angle of incidence   Angle of reflection

Incidence ray     Refracted ray

**b)**

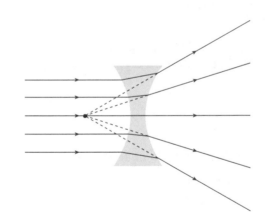

**6.** Magnetic; increase; potential difference; coil, reduced.

**7.** In a step-up transformer, there are more turns in the secondary coil than the primary coil so a greater potential difference leaves the secondary coil. In a step-down transformer there are fewer turns in the secondary coil than the primary coil so a smaller potential difference leaves the secondary coil.

**8.** Gravity attracts dust, rocks and gas. The mass increases and heat is created. Hydrogen fuses to form helium and a star is created.

# Notes

## Acknowledgements

The author and publisher would like to thank everyone who has contributed to this book:

p.10    ©iStockphoto.com / Peter Galbraith
p.30    ©iStockphoto.com / Stephen Sweet
p.67    ©iStockphoto.com

**ISBN 978-1-905896-40-0**

Published by Lonsdale, an imprint of HarperCollins*Publishers*.

©2007 Lonsdale.

**Author:** Andrew Catterall

**Project Editor:** Charlotte Christensen

**Cover Design:** Angela English

**Concept Design:** Sarah Duxbury and Helen Jacobs

**Designers:** Anne-Marie Taylor, Ian Wrigley and Paul Oates

**Artwork**: Lonsdale

Printed in the UK.

**Mixed Sources**
Product group from well-managed forests and other controlled sources
www.fsc.org   Cert no. SW-COC-001806
© 1996 Forest Stewardship Council

FSC is a non-profit international organisation established to promote the responsible management of the world's forests. Products carrying the FSC label are independently certified to assure consumers that they come from forests that are managed to meet the social, economic and ecological needs of present and future generations.

Find out more about HarperCollins and the environment at **www.harpercollins.co.uk/green**

## Author Information

Andrew Catterall worked as a science teacher, specialising in Physics, for 10 years before becoming a science consultant for an LEA. In his current role he works closely with the exam boards and has an excellent understanding of the new science specifications, which he is helping to implement in local schools.

The main objective in writing this book was to produce a user-friendly revision guide for students, which would also act as a useful reference for teachers. As such, this guide provides full coverage of all the essential material, cross-referenced to the specification for ease of use and presented in a clear and interesting manner that is accessible to everyone.

# Index